FROM MEASURES TO
ITÔ INTEGRALS

African Institute of Mathematics Library Series

The African Institute of Mathematical Sciences (AIMS), founded in 2003 in Muizenberg, South Africa, provides a one-year postgraduate course in mathematical sciences for students throughout the continent of Africa. The **AIMS LIBRARY SERIES** is a series of short innovative texts, suitable for self-study, on the mathematical sciences and their applications in the broadest sense.

AIMS Library Series

FROM MEASURES TO ITÔ INTEGRALS

EKKEHARD KOPP
University of Hull

CAMBRIDGE
UNIVERSITY PRESS

CAMBRIDGE UNIVERSITY PRESS
Cambridge, New York, Melbourne, Madrid, Cape Town,
Singapore, São Paulo, Delhi, Tokyo, Mexico City

Cambridge University Press
The Edinburgh Building, Cambridge CB2 8RU, UK

Published in the United States of America by Cambridge University Press, New York

www.cambridge.org
Information on this title: www.cambridge.org/9781107400863

First published 2011

Printed in the United Kingdom at the University Press, Cambridge

A catalogue record for this publication is available from the British Library

Library of Congress Cataloguing in Publication data
Kopp, Ekkehard (Peter Ekkehard), 1944–
From Measures to Itô Integrals / Ekkehard Kopp.
   p.   cm. – (African Institute of Mathematics Library Series)
Includes bibliographical references and index.
ISBN 978-1-107-40086-3 (pbk.)
1. Measure theory – Textbooks.   I. Title.
QA312.K5867   2011
515′.42–dc22

2010050362

ISBN 978-1-107-40086-3 Paperback

Contents

Preface

Undergraduate mathematics syllabi vary considerably in their coverage of measure-theoretic probability theory, so beginning graduates often find substantial gaps in their background when attending modules in advanced analysis, stochastic processes and applications. This text seeks to fill some of these gaps concisely. The exercises form an integral part of the text. The material arose from my experience of teaching AIMS students between 2004 and 2007, of which I retain many fond memories. The AIMS series format allows few explorations of byways; and the objective of arriving at a reasonably honest but concise account of the Itô integral decided most of the material. With motivation from elementary probability we discuss measures and integrals, leading via L^2-theory and conditional expectation to discrete martingales and an outline proof of the Radon–Nikodym Theorem. The last two chapters introduce Brownian Motion and Itô integrals, with a brief look at martingale calculus. Here proofs of several key results are only sketched briefly or omitted. The Black–Scholes option pricing model provides the main application. None of the results presented are new; any remaining errors are mine.

Three happy, lively AIMS student cohorts suffered my attempts to introduce them to mathematical finance. My sincere thanks go to them, to the AIMS team for their support, to Alan Beardon for his encouragement, to two helpful reviewers, to Marek Capinski for wise advice and above all to my wife Margaret for her patient, loving support.

1
Probability and measure

1.1 Do probabilists need measure theory?

Measure theory provides the theoretical framework essential for the development of modern probability theory. Much of elementary probability theory can be carried through with only passing reference to underlying sample spaces, but the modern theory relies heavily on measure theory, following *Kolmogorov's* axiomatic framework (1932) for probability spaces. The applications of *stochastic processes*, in particular, are now fundamental in physics, electronics, engineering, biology and finance, and within mathematics itself. For example, *Itô's stochastic calculus* for *Brownian Motion* (BM) and its extensions rely wholly on a thorough understanding of basic measure and integration theory. But even in much more elementary settings, effective choices of sample spaces and σ-fields bring advantages – good examples are the study of random walks and branching processes. (See [S], [W] for nice examples.)

1.2 Continuity of additive set functions

What do we mean by saying that we pick the number $x \in [0, 1]$ at random? 'Random' plausibly means that in each trial with uncertain outcomes, each outcome is 'equally likely' to be picked. Thus we seek to impose the *uniform probability distribution* on the set (or *sample space*) Ω of possible outcomes of an experiment. If Ω has n elements, this is trivial: for each outcome ω, the probability that ω occurs is $\frac{1}{n}$. But when $\Omega = [0, 1]$ the 'number' of possible choices of $x \in [0, 1]$ is infinite,

1

even uncountable. (Recall that the set $\mathbb{Q}$ of *rational* numbers is countable, while the set $\mathbb{R}$ of *real* numbers is uncountable.) We cannot define
the 'uniform probability' on $[0, 1]$ as a function of points x or singletons
$\{x\}$; however, we can first define our probability function or (*Lebesgue*)
measure m just for *intervals*: if $0 \le a < b \le 1$, we set $m([a, b]) = b - a$.
Thus measure, or 'probability', is a function of *sets*, not of points. The
challenge is to extend this idea to more general sets in $[0, 1]$.

With such an extension we can determine $m(\{x\})$ for fixed
$x \in [0, 1]$: for $\varepsilon > 0$, $\{x\} \subset \left[x - \frac{\varepsilon}{2}, x + \frac{\varepsilon}{2}\right]$, so if we assume m to
be *monotone* (i.e. $A \subset B$ implies $m(A) \le m(B)$), then we must con-
clude that $m(\{x\}) = 0$. On the other hand, since *some* number between
0 and 1 *is* chosen, it is *not impossible* that it could be our x. Thus a
non-empty set A can have $m(A) = 0$.

The probability that any one of a *countable* set of reals $A =
\{x_1, x_2, \ldots, x_n, \ldots\}$ is selected should also be 0, since for any $\varepsilon > 0$
we can *cover* each x_n by an interval $I_n = \left[x_n - \frac{\varepsilon}{2^{n+2}}, x_n + \frac{\varepsilon}{2^{n+2}}\right]$ so
that $A \subset \cup_{n=1}^{\infty} I_n$ with *total length* $\Sigma_{n=1}^{\infty} m(I_n) < \varepsilon$. We just need the
'obvious' property that $m(A) = \sum_{n=1}^{\infty} m(\{x_n\})$ for our conclusion.

We generalise this to demand the *countable additivity property*
of *any* probability function $A \mapsto P(A)$, i.e. if $(A_n)_{n \ge 1}$ are dis-
joint, then $P\left(\cup_{n=1}^{\infty} A_n\right) = \sum_{n=1}^{\infty} P(A_n)$. We shall formalise this in
Definition 1.9.

This demand looks very reasonable, and is an essential feature of
the calculus of probabilities. It implies *finite* additivity: if $A_1, \ldots, A_n$
($n \in \mathbb{N}$) are disjoint, then $P\left(\cup_{i=1}^{n} A_i\right) = \Sigma_{i=1}^{n} P(A_i)$. Simply
let $A_i = \emptyset$ for $i > n$ (see Proposition 1.3).

Remark 1.1 Our example suggests a useful description of the 'negli-
gible' (or *null*) sets in $[0, 1]$ (and by the same token in $\mathbb{R}$) for Lebesgue
measure m: the set A is m-*null* if for every $\varepsilon > 0$ there is a sequence
$(I_n)_{n \ge 1}$ of intervals of total length $\sum_{n=1}^{\infty} m(I_n) < \varepsilon$, so that $A \subset
\cup_{n=1}^{\infty} I_n$. (Note that the I_n need not be disjoint.) This requirement will
characterise sets $A \subset \mathbb{R}$ with Lebesgue measure $m(A) = 0$.

Example 1.2 The *Cantor set* provides an uncountable m-null set in
$[0, 1]$. Start with the interval $[0, 1]$, remove the interval $\left(\frac{1}{3}, \frac{2}{3}\right)$, obtaining
the set C_1, which consists of the two intervals $\left[0, \frac{1}{3}\right]$ and $\left[\frac{2}{3}, 1\right]$. Next
remove the 'middle thirds' $\left(\frac{1}{9}, \frac{2}{9}\right)$, $\left(\frac{7}{9}, \frac{8}{9}\right)$ of these two intervals, leaving

C_2, consisting of four intervals each of length $\frac{1}{9}$, etc. At the nth stage we have a set C_n, consisting of 2^n disjoint closed intervals, each of length $\frac{1}{3^n}$. Thus the total length of C_n is $\left(\frac{2}{3}\right)^n$, which goes to 0 as $n \to \infty$.

We call $C = \bigcap_{n=1}^{\infty} C_n$ the *Cantor set*, which is contained in each C_n, hence is m-null. Using ternary expansions, you may now show that C is uncountable (just as with decimal expansions for $[0, 1]$).

We develop some abstract probability theory. Any given set Ω can serve as *sample space*, and we consider ('future') *events* A, B from a given class $\mathcal{A}$ of subsets of Ω. We wish to define the probability $P(A)$ (resp. $P(B)$) as numbers in $[0, 1]$. Clearly, we would then also wish to know $P(A \cup B)$, $P(A^c)$, $P(A \cap B)$, etc. Thus the class $\mathcal{A}$ of sets on which P is defined should contain Ω (and $P(\Omega) = 1$) and together with A, B it should also contain $A \cup B$ and A^c. This ensures that $\mathcal{A}$ also contains $A \cap B$: $A^c \cup B^c = (A \cap B)^c$ is in $\mathcal{A}$, hence also $A \cap B$. Such a class $\mathcal{A}$ is a *field*. (This, now standard, use of the term 'field' in probability theory is somewhat unfortunate, and invites confusion with its usual algebraic meaning. Some authors seek to avoid this by using the term 'algebra' instead. We shall not do so.)

We demand that P is *additive,* i.e. for disjoint $A, B \in \mathcal{A}$ we have $P(A \cup B) = P(A) + P(B)$. This suffices for our first result.

Proposition 1.3 *If $A, B \in \mathcal{A}$ and $A \subset B$, then we have $P(B \backslash A) = P(B) - P(A)$. Hence $P(\emptyset) = 0$. Moreover, P is monotone: $A \subset B$ implies $P(A) \leq P(B)$.*

Proof $B \backslash A = B \cap A^c$, so $B \backslash A$ is in $\mathcal{A}$. But $B = A \cup (B \backslash A)$ and these sets are disjoint. Hence $P(B) = P(A) + P(B \backslash A)$. For the second claim, use $B = A$. The final claim follows as P is non-negative.

Exercise 1.4 Show that for any A, B in $\mathcal{A}$ (disjoint or not)

$$P(A \cup B) + P(A \cap B) = P(A) + P(B).$$

Given any probability P (see Definition 1.9 below), we call an *event* A *P-null* if $P(A) = 0$. An event B is called *almost sure* (or *full*) if $P(B) = 1$ (so that B^c is P null, since $P(B) + P(B^c) = P(\Omega)$). A property (e.g. of some function) holds *almost surely* if it holds on a full set (i.e. except possibly on some null set).

Example 1.5 (i) Let $\Omega = \{1, 2, \dots, n\}$ (or any finite set) $\mathcal{A} = 2^\Omega$ its *power set* (the class of all subsets) and $P(A) = \frac{\#A}{n}$, where $\#A$ is the number of points in A.

(ii) Let Ω be any set, $\mathcal{A} = 2^\Omega$, $\delta_x(A) = 1$ for $x \in A$, else 0. Then $P = \delta_x$ is the *point mass at* x (also called a *Dirac δ-measure*). Of course, the power set is always a field.

(iii) For finite or countable sample spaces, probability distributions can be built from Dirac measures: call a probability P on $\mathbb{R}$ *discrete* if there is a countable full subset C (i.e. $P(C) = 1$). This is obviously equivalent to P having the form $P = \sum_{i=1}^\infty p_i \delta_{x_i}$ for some real sequences $(x_i)_{i \geq 1}, (p_i)_{i \geq 1}$ with $p_i > 0$ and $\sum_{i=1}^\infty p_i = 1$.

The following distributions should be familiar:

(a) *Bernoulli:* $P = p\delta_1 + (1-p)\delta_0, 0 < p < 1$.

(b) *Binomial Bi(n,p):* $P = \sum_{i=1}^n p_i \delta_i$,
where $p_i = \binom{n}{i} p^i (1-p)^{n-i}, 0 \leq i \leq n, 0 < p < 1$.

(c) *Geometric Geo(p):* $P = \sum_{i=1}^\infty p_i \delta_i$,
where $p_i = p(1-p)^{i-1}, i \geq 1, 0 < p < 1$.

(d) *Negative binomial NegB(n,p):* $P = \sum_{i=n}^\infty p_i \delta_i$,
where $p_i = \binom{i-1}{n-1} p^i (1-p)^{i-n}, i \geq n, 0 < p < 1$.

(e) *Poisson Po(λ):* $P = \sum_{i=1}^\infty p_i \delta_i$,
where $\lambda > 0$ and $p_i = e^{-\lambda} \frac{\lambda^k}{i!}, i \geq 0$.

You may know these better as distributions of well-known classes of *random variables*.

Example 1.6 For a different example, we consider a field that enables us to generate Lebesgue measure on $\mathbb{R}$.

Let $\Omega = \mathbb{R}$. Consider left-open, right-closed intervals, i.e. of the form $(a, b]$ for $a, b \in [-\infty, \infty]$ (the set of *extended reals*, which consists of $\mathbb{R} \cup \{-\infty, \infty\}$) and where by convention we set $(a, \infty] = (a, \infty)$ for $-\infty \leq a \leq \infty$. We then define

$$\mathcal{A}_0 = \{\cup_{i=1}^n (a_i, b_i] : a_1 \leq b_1 \leq a_2 \leq \dots \leq b_n, \ n \geq 1\},$$

so that $\mathcal{A}_0$ is the class of all finite disjoint unions of such intervals. We define the measure of such a union as

$$m\left(\cup_{i=1}^n (a_i, b_i]\right) = \sum_{i=1}^n (b_i - a_i).$$

This fits with our earlier informal definition for closed intervals, since we showed that $m(\{x\}) = 0$ for any x and m is finitely additive.

Exercise 1.7 Verify that $\mathcal{A}_0$ is a field. Would this remain true if we had used open (or closed) intervals instead?

In practice, as can already be seen from examples (d) and (e) above, we are driven to considering unions and intersections of an *infinite sequence of events*. As another example, in an infinite sequence of coin tosses, what is the probability that 'heads' will occur infinitely often? We shall see that this depends crucially on our assumptions about the probability of success at each stage.

The *Borel–Cantelli (BC) Lemmas* are the archetype of this sort of result. To formulate the first lemma, suppose that $A_1, A_2, \ldots, A_n, \ldots$ is a sequence in $\mathcal{A}$ with $\sum_{n=1}^{\infty} P(A_n) < \infty$. How should we find the probability of the event

$$A_{\text{i.o.}} = \{\omega : \omega \in A_n \text{ for infinitely many } n\}?$$

(As with $A_{\text{i.o.}}$, we shall use 'i.o.' as an abbreviation for 'infinitely often' throughout.) We must ensure that $P(A_{\text{i.o.}})$ makes sense. If ω belongs to infinitely many A_n, then for each $m \geq 1$ there is at least one $n \geq m$ with $\omega \in A_n$. So $\omega \in \cup_{n \geq m} A_n$ for all m. Thus we need to define the probability of the union of infinitely many A_n if we are to get further. This leads first to:

Definition 1.8 A class $\mathcal{F}$ of subsets of a given set Ω is a σ-field (use of the term σ-algebra is also common) of subsets of Ω if:
(i) $\Omega \in \mathcal{F}$.
(ii) $A \in \mathcal{F}$ implies $A^c \in \mathcal{F}$.
(iii) $\{A_n : n \in \mathcal{N}\} \subset \mathcal{F}$ implies $\bigcup_{n=1}^{\infty} A_n \in \mathcal{F}$.
   Thus $\mathcal{F}$ is closed under complements and countable unions.

A field, and indeed any family $\mathcal{A}$ of subsets of Ω, *generates* a minimal σ-field $\sigma(\mathcal{A})$: we define

$$\sigma(\mathcal{A}) = \cap\{\mathcal{G} : \mathcal{G} \text{ is a } \sigma\text{-field}, \mathcal{A} \subset \mathcal{G}\}.$$

It is easily verified that $\sigma(\mathcal{A})$ satisfies (i)–(iii) of Definition 1.8 A key example is given by the *Borel σ-field* $\mathcal{B} = \sigma(\mathcal{A}_0)$ with the field $\mathcal{A}_0$

defined as in Example 1.6 above. In what follows, the pair $(\mathbb{R}, \mathcal{B})$ plays a central role.

For $\omega \in A_{\text{i.o.}}$, we need to have $\omega \in \cup_{n \geq m} A_n$ for *each* $m \geq 1$, so that we write

$$A_{\text{i.o.}} = \cap_{m \geq 1}(\cup_{n \geq m} A_n) := \limsup_{n \to \infty} A_n.$$

To see that $A_{\text{i.o.}}$ belongs to the σ-field $\mathcal{F}$, we show that $\mathcal{F}$ is also *closed under countable intersections.* Since $\mathcal{F}$ is closed under complements, if $(B_n)_{n \geq 1} \subset \mathcal{F}$ then each $B_n^c \in \mathcal{F}$, and by de Morgan's laws we have $(\cap_{n \geq 1} B_n)^c = \cup_{n \geq 1} B_n^c$, so that $\cap_{n \geq 1} B_n$ is the complement of a set in $\mathcal{F}$, hence is itself in $\mathcal{F}$. Thus we see that $A_{\text{i.o.}}$ is well-defined as soon as the sets A_n belong to a σ-field of sets in Ω.

But this still does not tell us how to find its probability.

Definition 1.9 A triple $(\Omega, \mathcal{F}, P)$ is a probability space if Ω is any set, $\mathcal{F}$ is a σ-field of subsets of Ω and the function $P : \mathcal{F} \to [0, 1]$ satisfies:
(i) $P(\Omega) = 1$.
(ii) $\{A_n : n \in N\} \subset \mathcal{F}$ and $A_n \cap A_m = \emptyset$ for $n \neq m$ imply that

$$P\left(\cup_{n=1}^{\infty} A_n\right) = \sum_{n=1}^{\infty} P(A_n).$$

We say that the *probability* P is a σ-*additive* (also called *countably additive*) set function.

We may equally define $\liminf_{n \to \infty} A_n = \cup_{n \geq 1}(\cap A_{m \geq n})$. This set contains all points that *eventually* belong to sets in the sequence $(A_n)_n$. We also write A_{ev} for this set.

Exercise 1.10 Check: $(\liminf_{n \to \infty} A_n)^c = \limsup_{n \to \infty} A_n^c$. Compare this with a similar result for the upper and lower limits of a sequence of real numbers – recall that for a real sequence (a_n) we define $\limsup_n a_n$ as $\inf_{n \geq 1}(\sup_{m \geq n} a_m)$ and $\liminf_n a_n$ as $\sup_{n \geq 1}(\inf_{m \geq n} a_m)$. You should prove that (a_n) converges if and only if (iff) these quantities coincide! We use this fact in later chapters.

Any countable union $\bigcup_{k=1}^{\infty} A_k$ can be written as a *disjoint* union: let $B_k = A_k \backslash \bigcup_{j=1}^{k-1} A_j$, then clearly $B_k \subset A_k$ for each k and $B_j \cap B_k = \emptyset$ when $j \neq k$. You should verify that $\bigcup_{k=1}^{\infty} B_k = \bigcup_{k=1}^{\infty} A_k$.

Exercise 1.11 Prove that a finitely additive set function $P : \mathcal{F} \to [0, 1]$ is σ-additive iff the following statement holds: whenever $(B_n)_{n\geq 1}$ in $\mathcal{F}$ decreases to the empty set ($\cap_{n\geq 1} B_n = \emptyset$), then $\lim_{n\to\infty} P(B_n) = 0$. (This is often called '*continuity*' of P at $\emptyset$.)

To find $P(\limsup_{n\to\infty} A_n)$, we first need some more simple consequences of the definitions:

Proposition 1.12 *Let* $(\Omega, \mathcal{F}, P)$ *be a probability space.*
(i) If $(A_i)_{i\geq 1}$ *in* $\mathcal{F}$, *then* $P(\cup_{i\geq 1} A_i) \leq \sum_{i=1}^{\infty} P(A_i)$.
(ii) If $A_i \subset A_{i+1}$, $(i \geq 1)$, *then* $P(\cup_{i\geq 1} A_i) = \lim_{n\to\infty} P(A_n)$.
(iii) If $A_{i+1} \subset A_i$, $(i \geq 1)$, *then* $P(\cap_{i\geq 1} A_i) = \lim_{n\to\infty} P(A_n)$.

Exercise 1.13 Prove Proposition 1.12 and show that (ii) and (iii) are equivalent ways of formulating σ-additivity of additive P. (Here (i) says that P is *countably subadditive*.)

We introduce notation for the *convergence of sets*: write $A_n \to A$ if $\limsup_n A_n = \liminf_n A_n = A$ (alternatively, $A_{\text{i.o.}} = A_{\text{ev}} = A$). Note the analogy with convergent sequences! As a special case, we write $A_n \uparrow A$ if $A_n \subset A_{n+1}$ for all n and $A = \cup_{n=1}^{\infty} A_n$. Similarly, $A_n \downarrow A$ if $A_{n+1} \subset A_n$ for all n and $A = \cap_{n=1}^{\infty} A_n$.

Proposition 1.14 *If* $A_n \to A$, *then* $P(A_n) \to P(A)$.

Proof If $A_n \to A$, then $A = A_{\text{i.o.}} = A_{\text{ev}}$, so $P(A) = P(A_{\text{i.o.}}) = P(A_{\text{ev}})$. We need to show that

$$\limsup_n P(A_n) = \liminf_n P(A_n) = P(A).$$

But $\liminf_n P(A_n) \leq \limsup_n P(A_n)$ always holds, so we just need to show that $P(\liminf_n(A_n)) \leq \liminf_n P(A_n)$. Now for each $k \geq 1$

$$\cap_{n\geq k} A_n \subset A_k, \quad \text{hence } P(\cap_{n\geq k} A_n) \leq P(A_k)$$

and the result follows by Proposition 1.12 (ii) on letting $k \to \infty$, since $\cap_{n\geq k} A_n \uparrow \liminf_n A_k = A$.

Exercise 1.15 Show that $\limsup_n P(A_n) \leq P(\limsup_n A_n)$.

For the first BC Lemma, let $B_m = \cup_{n \geq m} A_n$ so that (B_m) decreases, hence by 1.12(ii) and 1.12(i)

$$P(\cap_{m \geq 1} B_m) = \lim_m P(B_m) = \lim_m P(A_m \cup A_{m+1} \cup \ldots)$$
$$\leq \lim_m (P(A_m) + P(A_{m+1}) + \ldots) = 0$$

since the series (of real numbers!) $\sum_{n=1}^{\infty} P(A_n)$ converges.

We have proved:

Lemma 1.16   *(First Borel–Cantelli (BC) Lemma): If $(A_n)_n$ is a sequence of events with $\sum_{n=1}^{\infty} P(A_n) < \infty$, then $P(\limsup_n A_n) = 0$.*

Thus, if we have a sequence of events whose probabilities decrease quickly enough to keep their sum finite (for example, if $P(A_{n+1}) = 0.999 P(A_n)$ for each n), then it is *certain* (i.e. the probability is 1) that only finitely many of them will occur. This may not be unduly surprising, but it did need a proof.

1.3 Independent events

The first BC Lemma is immediate from our definitions. Matters are very different, however, when the series $\sum_{n=1}^{\infty} P(A_n)$ diverges. Then we do have a *second BC Lemma*, but this applies only when the sequence of events $(A_n)_n$ is *independent*. Recall some basic definitions:

Definition 1.17   Let $(\Omega, \mathcal{F}, P)$ be a probability space. For $A, B \in \mathcal{F}$ with $P(B) > 0$, define

$$P(A|B) = \frac{P(A \cap B)}{P(B)}$$

as the *conditional probability* of A, *given* B.

Exercise 1.18   Verify that the function $P_B : A \to P(A|B)$ is again a probability. (*Hint*: if $(A_n)_n$ are pairwise disjoint sets in $\mathcal{F}$, then $(A_n \cap B)_n$ are also pairwise disjoint.)

Exercise 1.19 Suppose that $A, B_n \in \mathcal{F}$ with $(B_n)_n$ pairwise disjoint, $P(B_n) \neq 0$ for all n and $\cup_{n=1}^{\infty} B_n = \Omega$. Prove that

$$P(A) = \sum_{n=1}^{\infty} P(A|B_n)P(B_n).$$

This is often called the *Theorem of Total Probability*.

Definition 1.20 Events A, B in $\mathcal{F}$ are *independent* if

$$P(A \cap B) = P(A)P(B).$$

When $P(B) > 0$, this is the same as the more natural requirement that B should 'have no influence' on A; i.e. $P(A|B) = P(A)$. However, our definition still makes sense if $P(B) = 0$.

Care must be taken when generalising this definition to three or more sets (e.g. A, B, C): it is not enough simply to require that $P(A \cap B \cap C) = P(A)P(B)P(C)$.

Exercise 1.21 Find examples of sets in $\mathbb{R}$ to justify this claim.

As our general definition of independence of events, we therefore require:

Definition 1.22 Let $(\Omega, \mathcal{F}, P)$ be a probability space. Events $A_1, A_2, \ldots, A_n$ in $\mathcal{F}$ are *independent* if for each choice of indices $1 \leq i_1 < i_2 < \ldots < i_k \leq n$

$$P\left(\cap_{m=1}^{k} A_{i_m}\right) = P(A_{i_1})P(A_{i_2}) \ldots P(A_{i_k}) = \prod_{m=1}^{k} P(A_{i_m}).$$

A sequence $(A_n)_{n \geq 1}$ of events (or any family $(A_\alpha)_\alpha$) is *independent* if every finite subset $A_{i_1}, A_{i_2}, \ldots, A_{i_k}$ of events is independent.

With this machinery we formulate:

Lemma 1.23 *(Second Borel–Cantelli (BC) Lemma): If the sequence $(A_n)_{n \geq 1}$ is independent and $\sum_{n=1}^{\infty} P(A_n) = \infty$, then $P(\limsup_n A_n) = 1$.*

Proof To prove that $P\left(\cap_{k=1}^{\infty}\left(\cup_{n=k}^{\infty}A_n\right)\right) = 1$ it will suffice to show that for each $k \geq 1$, $P\left(\cup_{n=k}^{\infty}A_n\right) = 1$. This follows from Proposition 1.12 (iii): $\cup_{n=k}^{\infty}A_n$ decreases as k increases, so that

$$\lim_{k\to\infty} P\left(\cup_{n=k}^{\infty}A_n\right) = P\left(\cap_{k=1}^{\infty}\left(\cup_{n=k}^{\infty}A_n\right)\right).$$

Now consider $\cap_{n=k}^{m}A_n^c$ for a fixed $m > k$. By de Morgan's laws, we have $\left(\cup_{n=k}^{m}A_n\right)^c = \cap_{n=k}^{m}A_n^c$. The (A_n^c) are also independent (check this yourself!), so for $k \geq 1$

$$P\left(\cap_{n=k}^{m}A_n^c\right) = \prod_{n=k}^{m} P\left(A_n^c\right) = \prod_{n=k}^{m}[1 - P(A_n)].$$

For $x \geq 0$, we have $1 - x \leq e^{-x}$ (use the Taylor series, or simply the derivative), so

$$\prod_{n=k}^{m}[1 - P(A_n)] \leq \prod_{n=k}^{m} e^{-P(A_n)} = e^{-\sum_{n=k}^{m} P(A_n)}.$$

Now recall that we have assumed that the series $\sum_n P(A_n)$ diverges. Hence for fixed k the partial sums $\sum_{n=k}^{m} P(A_n)$ grow beyond all bounds as $m \to \infty$. So, as $m \to \infty$ the right-hand side (RHS) of the inequality becomes arbitrarily small.

This proves that

$$1 - P\left(\cup_{n=k}^{m}A_n\right) = P\left(\cap_{n=k}^{m}A_n^c\right) \to 0 \text{ as } m \to \infty.$$

Now write $B_m = \cup_{n=k}^{m}A_n$. The sequence $(B_m)_m$ is increasing in m and its union is $\cup_{n=k}^{\infty}A_n$. Applying Proposition 1.12 (ii), we have

$$P\left(\cup_{n=k}^{\infty}A_n\right) = \lim_{m\to\infty} P(B_m) = 1$$

and the proof is complete.

1.4 Simple random walk

The following famous example illustrates the power of the BC Lemmas. (i) On the line, imagine a drunkard describing a symmetric random walk from 0, i.e. who at each step is equally likely to go left or right. How often does such a walk return to the starting point? The position reached after n steps is given by $S_n = \sum_{i=1}^{n} X_i$, where the X_i are independent Bernoulli random variables with

$$P(X_i = 1) = \frac{1}{2} = P(X_i = -1).$$

Note that $P(S_{2n-1} = 0) = 0$ for all n (Why?).

To find $P(S_{2n} = 0)$ we use the well-known approximation of $n!$ given by *Stirling's formula*

$$n! \approx \left(\frac{n}{e}\right)^n \sqrt{2\pi n} \text{ for large } n.$$

To get back to 0 in $2n$ steps, the drunkard has to have taken n steps to the right and n to the left. Thus the probability of return to 0 is (for large n)

$$P(S_{2n} = 0) = \binom{2n}{n}\left(\frac{1}{2}\right)^n\left(\frac{1}{2}\right)^{2n-n} = \frac{(2n)!}{(n!)^2}\left(\frac{1}{2}\right)^{2n}$$

$$\approx \frac{\left(\frac{2n}{e}\right)^{2n}\sqrt{2\pi(2n)}}{\left[\left(\frac{n}{e}\right)^n\sqrt{2\pi n}\right]^2}\left(\frac{1}{2}\right)^{2n} = \frac{1}{\sqrt{n\pi}}.$$

Now take $A_n = (S_{2n} = 0)$ in the second BC Lemma: the series $\sum_{n=1}^{\infty} P(A_n)$ diverges, as for large n we have $P(A_n) \approx \frac{1}{\sqrt{n\pi}}$ and $\sum_n \frac{1}{\sqrt{n}}$ is divergent. Thus we have shown that $P(S_{2n} = 0$ infinitely often$) = 1$.

(ii) In two dimensions, a similar 'walk' would take place on a grid, so there are four choices at each step, and we can regard the path as the composition of two one-dimensional excursions similar to the above. We can also take them as independent (this needs justification, which we omit), so that $P(S_{2n} = 0) \approx \frac{1}{n\pi}$. The series $\sum_{n=1}^{\infty} P(A_n)$ also diverges, since $\sum_n \frac{1}{n}$ diverges. Thus we again find that $P(S_{2n} = 0$ i.o.$) = 1$.

(iii) In three dimensions, similarly, $P(S_{2n} = 0) \approx \frac{1}{(n\pi)^{3/2}}$. But now the series $\sum_{n=1}^{\infty} P(A_n)$ converges by comparison with the convergent series $\sum_n \frac{1}{n^{3/2}}$. Thus the first BC Lemma shows that $P(S_{2n} = 0$ i.o.$) = 0$ for each n.

2

Measures and distribution functions

The next definition is motivated by Lebesgue measure, and extends that of probability spaces by relaxing the condition that $\mu(\Omega) = 1$. We now stipulate as part of the definition that the empty set has measure 0.

Definition 2.1 A *measurable space* $(\Omega, \mathcal{F})$ consists of a set Ω and a σ-field $\mathcal{F}$ of its subsets. The sets in $\mathcal{F}$ are called *measurable*. A *measure* on $(\Omega, \mathcal{F})$ is a function $\mu : \mathcal{F} \to [0, \infty]$ such that $\mu(\emptyset) = 0$ and for any disjoint sequence $(A_n)_n$ in $\mathcal{F}$, $\mu(\cup_{n=1}^{\infty} A_n) = \sum_{n=1}^{\infty} \mu(A_n)$. (This identity is expressed by saying that μ is *countably additive*.)

However, if there is a measurable set A with $\mu(A) < \infty$, we can employ additivity: with $B = A$ in Proposition 1.3 (and the same proof) we see again that $\mu(\emptyset) = \mu(A) - \mu(A) = 0$. Most of Proposition 1.12 also remains true with P replaced by μ: the only change is that in 1.12(iii) we need to demand that $\mu(A_1) < \infty$ for the arguments to make sense.

2.1 σ-finite measures

There are measure spaces where no non-empty set has finite measure! Setting $\mu(A) = +\infty$ for each non-empty set in the field $\mathcal{A}_0$, described in Example 1.6, and using Theorem 2.8 below yield such a measure on the Borel σ-field $\mathcal{B}$. To avoid such examples, we restrict our attention to:

Definition 2.2 A measure μ on a measurable space $(\Omega, \mathcal{F})$ is *finite* if $\mu(\Omega) < \infty$. It is *σ-finite* if there is a sequence $(F_n)_n$ in $\mathcal{F}$ with $\mu(F_n) < \infty$ for all n and $\cup_n F_n = \Omega$.

Recall that in a probability space $(\Omega, \mathcal{F}, P)$ we say that a certain property (e.g. of functions) holds *almost surely* (abbreviated as a.s.(P), or simply a.s.) if it holds on a full set (i.e. one whose complement is null). For a general measure μ, the term 'almost everywhere' (abbreviated as a.e.(μ), sometimes as μ-a.e., or just a.e.) is commonly used for the same concept – 'a.s.' will indicate that the measure in question is a probability measure.

Example 2.3 For $A = \cup_{i=1}^n (a_i, b_i] \in \mathcal{A}_0$ (see Example 1.6), length was given in an obvious way: $m(A) = \sum_{i=1}^n (b_i - a_i)$. Clearly, the sequence $F_n = (-n, n]$ shows that m is σ-finite. To extend this measure to more general sets, the first step is very natural: for $E \subset \mathbb{R}$, define *Lebesgue outer measure* m^* by

$$m^*(E) = \inf \left\{ \sum_{i=1}^{\infty} (b_i - a_i) : E \subset \cup_{i=1}^{\infty}(a_i, b_i] \right\}.$$

Note that by its definition, m^* is *monotone*: $A \subset B$ implies that $m^*(A) \leq m^*(B)$. We leave another property as an exercise: show that m^* is *countably subadditive*, i.e. that for any $(E_i)_i$

$$m^* \left(\cup_{i=1}^{\infty} E_i \right) \leq \sum_{i=1}^{\infty} m^*(E_i).$$

However, it is too much to expect that m^* is σ-*additive* on all subsets of $\mathbb{R}$; at least if we accept the *Axiom of Choice*. Informally, this axiom says that, given any collection of non-empty sets, we can form a new set by choosing one element from each set in the collection. The axiom is independent of the axioms of (Zermelo–Fraenkel) set theory, and we shall (like most mathematicians) accept its validity. With it we can construct a sequence of sets in $\mathbb{R}$ on which m^* is not σ-additive. First, we need a simple fact about m^* whose proof is left to you. Write $E + x = \{e + x : e \in E\}$ for $E \subset \mathbb{R}$ and $x \in \mathbb{R}$. Show that m^* is translation-invariant: $m^*(E + x) = m^*(E)$ for $E \subset \mathbb{R}$ and $x \in \mathbb{R}$.

Example 2.4 Define an equivalence relation on $[0, 1]$ by saying that $x \sim y$ iff $x - y \in \mathbb{Q}$. From each equivalence class (how many classes are there?), pick one element and call the set formed by these elements A. By the above, $m^*(A + q) = m^*(A)$ for any rational q. Also, if $q \neq r$, then $A + q$ and $A + r$ are disjoint, as is easily seen.

Suppose m^* is σ-additive. With $B = \cup_{q \in [-1,1] \cap \mathbb{Q}} (A + q)$, we have $m^*(B) = \sum_{q \in [-1,1] \cap \mathbb{Q}} m^*(A + q)$. But we have $[0,1] \subset B \subset [-1,2]$, so, by subadditivity of m^*, the first inclusion gives $1 \leq \sum_{q \in [-1,1] \cap \mathbb{Q}} m^*(A + q)$, hence $m^*(A) > 0$. On the other hand, the second inclusion tells us that $\sum_{q \in [-1,1] \cap \mathbb{Q}} m^*(A + q) \leq 2$ and hence $m^*(A) = 0$. The contradiction shows that m^* is not σ-additive over 2^Ω.

On the other hand, we will see shortly that m^* *is* a measure on the Borel σ-field $\mathcal{B} = \sigma(\mathcal{A}_0)$. Since $(a, b) = \cup_{n=k}^\infty \left(a, b - \frac{1}{n} \right]$ (when $\frac{1}{k} < b - a$), it follows that every open interval (and hence every open set, as it can be written as a *countable* union of open intervals) belongs to $\mathcal{B}$. So every closed set also belongs to $\mathcal{B}$.

Thus $\mathcal{B}$ could equally well be defined as the smallest σ-field containing all open sets (or all closed sets, or all intervals), since, conversely, any interval of the form $(a, b] = \cap_{n=1}^\infty \left(a, b + \frac{1}{n} \right)$ would then also belong to this σ-field. The flexibility we have in the type of interval we use is helpful in many situations. We started with half-open intervals because we want their lengths to add up correctly, with no gaps or overlaps: to ensure that length m is additive, we need in particular for $c \in (a, b)$ that $m((a, b]) = b - a = (b - c) + (c - a) = m((c, b]) + m((a, c])$.

We extend things a little further. A set N is m^*-*null* if $m^*(N) = 0$ (cf. also Remark 1.1). We write $\mathcal{N}$ for the collection of all m^*-null sets (note that subsets of null sets are null as m^* is monotone). The σ-field $\mathcal{L}$ is the σ-field generated by $\mathcal{B} \cup \mathcal{N}$. Sets in $\mathcal{L}$ will be called *Lebesgue-measurable*.

To show that m extends to a measure on $(\mathbb{R}, \mathcal{B})$, we quote without proof a key general result on the extension of measures. First we need some more definitions:

Definition 2.5 (i) $\mu^*\!:2^\Omega \to [0, \infty]$ is an *outer measure* if it is countably subadditive, monotone and $\mu^*(\emptyset) = 0$.

(ii) Let μ^* be an outer measure. $A \subset \Omega$ is μ^*-*measurable* if

$$\mu^*(E) = \mu^*(E \cap A) + \mu^*(E \backslash A)$$

for each $E \subset \Omega$.

It is easy to see that $m^* : 2^{\mathbb{R}} \to [0, \infty]$, defined in Example 2.3, is an outer measure, as is μ^*, defined in the Key Extension Theorem, which we state without proof:

Theorem 2.6 *(Caratheodory) Suppose $\mathcal{F}_0$ is a field of subsets of a set Ω and $\mu : \mathcal{F}_0 \to [0, \infty]$ is a measure. Define $\mu^*(E) = \inf \{ \sum_{i=1}^{\infty} \mu(A_i) : A_i \in \mathcal{F}_0, \ E \subset \cup_{i=1}^{\infty} A_i \}$ for $E \subset \Omega$. Then μ^* and μ agree on $\mathcal{F}_0$, every set in $\mathcal{F}_0$ is μ^*-measurable, and if μ is σ-finite, then μ^* is the unique extension of μ to a measure on the σ-field $\mathcal{F} = \sigma(\mathcal{F}_0)$.*

This reduces the proof of the existence of Lebesgue measure $m : \mathcal{L} \to [0, \infty]$ to two steps, which are also given without proof. The first applies to outer measures in general; the second ensures that real intervals are m^*-measurable:

(1) For any outer measure μ^*, the collection $\mathcal{M}(\mu^*)$ of μ^*-measurable sets is a σ-field containing all μ^*-null sets, and the restriction of μ^* to $\mathcal{M}(\mu^*)$ is a measure.

(2) If for a disjoint sequence $(A_i)_{i \geq 1}$ in $\mathcal{A}_0$ their union $A = \cup_{i=1}^{\infty} A_i$ also belongs to $\mathcal{A}_0$, then $m(A) = \sum_{i=1}^{\infty} m(A_i)$.

Thus Carathedory's Theorem ensures that sets in $\mathcal{A}_0$ are m^*-measurable, and that $m^* : \mathcal{L} \to [0, \infty]$ is the unique measure extending m, since $\mathcal{L} \subset \mathcal{M}(m^*)$. We write m instead of m^* for *Lebesgue measure on $\mathbb{R}$*. This measure is *complete*, i.e. subsets of m-null sets are measurable.

Finally, note that Example 2.4 provides us with a set A which is not in $\mathcal{L}$: if A were Lebesgue-measurable, then m^* would be σ-additive over the sequence $A + q : q \in \mathbb{Q}$. Thus we are justified in calling A a *non-measurable set* for Lebesgue measure.

2.2 Generated σ-fields and π-systems

Although a measure needs to be defined on a σ-field, we rarely have situations where these are given in advance. For example, in building Lebesgue measure we started with intervals, or, more precisely, with the field $\mathcal{A}_0$, and generated the Borel σ-field $\mathcal{B}$ from it. The question arises: under which conditions on a class $\mathcal{C}$ will the extension of a measure from $\mathcal{C}$ to $\sigma(\mathcal{C})$ be unique? It turns out, remarkably, that all we need is that our generating class should contain $\emptyset$ and be *closed under finite intersections*, i.e. if $A, B \in \mathcal{C}$, then $A \cap B \in \mathcal{C}$. We call such a class a π-*system*. Clearly, any field is a π-system. For the field $\mathcal{A}_0$, we have: if $a < b < c < d$, then $(a, c] \cap (b, d] = (b, c]$.

We need one further notion: a system $\mathcal{D}$ of subsets of Ω is a d-*system* (sometimes called a *monotone class*) if it contains Ω and is closed under (proper) difference (if $A \subset B$ in $\mathcal{D}$, then $B \backslash A \in \mathcal{D}$) and increasing countable unions (if $A_i \subset A_{i+1}, A_i \in \mathcal{D}$ for all $i \geq 1$, then $\cup_{i=1}^{\infty} A_i \in \mathcal{D}$). It is usually easier to check these requirements than those for σ-fields. As for σ-fields, we write $d(\mathcal{C})$ for the smallest d-system containing $\mathcal{C}$ – note that this is again the intersection of all d-systems containing $\mathcal{C}$ – and call it the d-system *generated by* $\mathcal{C}$. The link with σ-fields is clear from:

Exercise 2.7 Show that a collection $\mathcal{C}$ of subsets of Ω is a σ-field iff it is both a π-system and a d-system.

Theorem 2.8   *Monotone Class Theorem for π-systems. For any π-system $\mathcal{C}$ on Ω, we have:* $d(\mathcal{C}) = \sigma(\mathcal{C})$.

Proof Define $\mathcal{C}_1 = \{B \in d(\mathcal{C}) : B \cap C \in d(\mathcal{C})$ for all C in $\mathcal{C}\}$. This condition holds whenever $B \in \mathcal{C}$ as $\mathcal{C}$ is a π-system. We verify that $\mathcal{C}_1$ is a d-system: first, $\Omega \in \mathcal{C}_1$. If $A \subset B$ are in $\mathcal{C}_1$, then for $C \in \mathcal{C}, (B \backslash A) \cap C = (B \cap C) \backslash (A \cap C)$ is in $d(\mathcal{C})$ by the definition of d-systems. Thus $B \backslash A \in \mathcal{C}_1$. Finally, let $B_n \uparrow B$ with $B_n \in \mathcal{C}_1$ for all n. Thus for any $C \in \mathcal{C}, B_n \cap C \uparrow B \cap C$ and so $B \cap C \in d(\mathcal{C})$. Hence $B \in \mathcal{C}_1$, as required. But $\mathcal{C}_1$ contains $\mathcal{C}$ and hence it equals $d(\mathcal{C})$. Next, define $\mathcal{C}_2 = \{A \in d(\mathcal{C}) : A \cap B \in d(\mathcal{C})$ for all B in $\mathcal{C}_1\}$. We have just seen that this collection contains $\mathcal{C}$. But it is again a d-system (just as for $\mathcal{C}_1$) and thus equals $d(\mathcal{C})$. This means that $d(\mathcal{C})$ is a π-system. Hence by Exercise 2.7 it must equal $\sigma(\mathcal{C})$.

Thus, if we begin with an π-system of subsets of Ω on which a certain property is true, then we only need to establish that the class $\mathcal{C}$ of subsets on which the property holds is a d-system. For then the property holds on $d(\mathcal{C}) = \sigma(\mathcal{C})$. Our first application is to show how this applies to σ-finite measures.

Theorem 2.9 *Suppose $\mathcal{F}$ is a σ-field on Ω, $\mathcal{C}$ is a π-system with $\sigma(\mathcal{C}) = \mathcal{F}$, and the σ-finite measures P, Q agree on $\mathcal{C}$. Then they agree on $\mathcal{F}$.*

Proof Let $\mathcal{G}$ be the class of subsets $\mathcal{G} = \{A \in \mathcal{F} : P(A) = Q(A)\}$. We are given that $P = Q$ on $\mathcal{C}$ so $\mathcal{C} \subset \mathcal{G}$. Thus by Theorem 2.8 we just need to verify that $\mathcal{G}$ is a d-system. This is trivial: Ω is in $\mathcal{G}$ by hypothesis; for $A \subset B$ in $\mathcal{G}$, we have $P(B \backslash A) = P(B) - P(A) = Q(B) - Q(A) = Q(B \backslash A)$, again using Proposition 1.3 (which applies to σ-finite measures). Finally, for an increasing sequence in $\mathcal{G}$ with $A_n \uparrow A$ we have $P(A) = \lim_n P(A_n) = \lim_n Q(A_n) = Q(A)$.

2.3 Distribution functions

Theorem 2.9 shows in particular that *Lebesgue measure m* is the unique extension of m from $\mathcal{A}_0$ to $\mathcal{B}$. More generally, any σ-finite measure, and so every probability P, on $(\mathbb{R}, \mathcal{B})$ is determined by its values on intervals of the form $(a, b]$; in fact, by intervals of the form $(-\infty, x]$. This enables us to define the *distribution function* of the probability P as the function $F_P : \mathbb{R} \to [0, 1]$ given for $x \in \mathbb{R}$ by $F_P(x) = P((-\infty, x])$. The most natural examples are distribution functions of random variables (see Chapter 3): for $X : \Omega \to \mathbb{R}$ we define the probability $P_X : \mathcal{B} \to [0, 1]$ by $P_X(A) = P(X \in A) = P \circ X^{-1}(A)$. P_X is often called the *law* of the random variable X. Then we write F_X for F_{P_X}, so that $F_X(x) = P(X \leq x)$.

Exercise 2.10 Check that P_X is a measure on $(\mathbb{R}, \mathcal{B}(\mathbb{R}))$.

The probability P on $(\mathbb{R}, \mathcal{B})$ determines the distribution function F_P. The converse also holds:

Proposition 2.11 *If $F_P = F_Q$, then $P = Q$. In particular, for random variables X, Y, if $F_X = F_Y$, then $P_X = P_Y$.*

Proof By the previous theorem, we only need to show that $P = Q$ on intervals $(a, b]$. Since $(a, b] = (-\infty, b) \backslash (-\infty, a]$

$$P((a,b]) = P((-\infty,b]) - P((-\infty,a]) = F_P(b) - F_P(a)$$
$$= F_Q(b) - F_Q(a) = Q((-\infty,b)) - Q((-\infty,a])$$
$$= Q((a,b]).$$

The main properties of distribution functions are left as

Exercise 2.12 If P is a probability on $\mathbb{R}$, verify that its distribution function F_P is increasing, has right and left limits at all points and is right-continuous.

In fact, any finite, non-decreasing and right-continuous function real F (also called a distribution function) will determine a measure μ_F on $(\mathbb{R}, \mathcal{B})$ by first setting $\mu_F((a,b]) = F(b) - F(a)$ for intervals of this form, and then extending to the Borel σ-field $\mathcal{B}$. Clearly, two such functions differing by a constant define the same measure. Such *Lebesgue–Stieltjes measures* μ_F assign finite values to all finite intervals. It can be shown (see e.g. [L]) that there is a one–one correspondence between (equivalence classes of) distribution functions and Lebesgue–Stieltjes measures.

3

Measurable functions/random variables

A function between measurable spaces, $f : (\Omega, \mathcal{F}) \rightarrow (\Omega', \mathcal{F}')$ is *measurable* if $f^{-1}(E') \in \mathcal{F}$ for all E' in $\mathcal{F}'$. (Recall that a map $f : (X, \mathcal{T}) \rightarrow (X', \mathcal{T}')$ between topological spaces is *continuous* if $f^{-1}(E') \in \mathcal{T}$ for all E' in $\mathcal{T}'$.) The *inverse image* of a set A under a function f is very well-behaved with regard to operations on sets:

Exercise 3.1 Suppose Ω, Ω' are sets and $f : \Omega \rightarrow \Omega'$ is a function. Write $f^{-1}(A) = \{\omega \in \Omega : f(\omega) \in A\}$ for any $A \subset \Omega'$. Show that:
(i) if $A \subset B$, then $f^{-1}(A) \subset f^{-1}(B)$;
(ii) for any family $(A_\alpha)_{\alpha \in \Lambda}$ of subsets of Ω'

$$f^{-1}(\cup_{\alpha \in \Lambda} A_\alpha) = \cup_{\alpha \in \Lambda} f^{-1}(A_\alpha),$$
$$f^{-1}(\cap_{\alpha \in \Lambda} A_\alpha) = \cap_{\alpha \in \Lambda} f^{-1}(A_\alpha);$$

(iii) $f^{-1}(A^c) = (f^{-1}(A))^c$;
(iv) if $A \cap B = \emptyset$, then $f^{-1}(A) \cap f^{-1}(B) = \emptyset$.
Which, if any, of statements (i)–(iv) hold with f^{-1} replaced by f?

For measurable $f : (\Omega, \mathcal{F}) \rightarrow (\Omega', \mathcal{F}')$, the collection of inverse images $\{f^{-1}(F') : F' \in \mathcal{F}'\}$ is thus again a σ-field, and is obviously contained in $\mathcal{F}$. This is the σ-field *generated by f*, denoted by $f^{-1}(\mathcal{F}')$.

3.1 Measurable real-valued functions

We work with *real-valued* functions, where the range space is $(\mathbb{R}, \mathcal{B})$. Because the Borel σ-field $\mathcal{B}$ is generated by the intervals in $\mathbb{R}$, we can rephrase our definition in this case as follows:

Definition 3.2 A real-valued function $f : \Omega \to \mathbb{R}$ on the measurable space $(\Omega, \mathcal{F})$ is *$\mathcal{F}$-measurable* if

$$f^{-1}(I) = \{\omega \in \Omega : f(\omega) \in I\} \in \mathcal{F} \text{ for each real interval } I.$$

We use the flexibility of the Borel σ-field to express Definition 3.2 in different equivalent forms.

Exercise 3.3 Show that $f : \Omega \to \mathbb{R}$ is measurable iff any of the following hold:
(i) $f^{-1}((a, \infty)) \in \mathcal{F}$ for all real a;
(ii) $f^{-1}([a, \infty)) \in \mathcal{F}$ for all real a;
(iii) $f^{-1}((-\infty, a)) \in \mathcal{F}$ for all real a;
(iv) $f^{-1}((-\infty, a]) \in \mathcal{F}$ for all real a.

By convention, when $(\Omega, \mathcal{F}, P)$ is a probability space, measurable functions $\Omega \to \mathbb{R}$ are called *random variables* (often abbreviated to r.v.) and are usually denoted by X, Y, Z, instead of f, g, h.

Exercise 3.4 Let $(\Omega, \mathcal{F}, P)$ be a probability space and let $X : \Omega \to \mathbb{R}$ be a random variable. Describe the σ-field $X^{-1}(\mathcal{B})$ generated by X when:
(i) X is constant;
(ii) $X = \mathbf{1}_B$ for some Borel set B;
(iii) X takes finitely many different values.

3.2 Lebesgue- and Borel-measurable functions

We take a brief look at the special case of functions $f : \mathbb{R} \to \mathbb{R}$. We say that f is *Lebesgue-measurable* if $f^{-1}(B) \in \mathcal{L}$ for every Borel set B. A stronger demand is that it is *Borel-measurable*, which means that $f^{-1}(B) \in \mathcal{B}$ for every Borel set B. There are many Borel-measurable functions as the next Exercise shows.

Exercise 3.5
(i) Suppose $f : \mathbb{R} \to \mathbb{R}$ is continuous. Show that f is Borel-measurable. Extend the definitions and this result to the case $f : \Omega \to \mathbb{R}$ where Ω is a metric space and the σ-field $\mathcal{F}$ contains all open sets.
(ii) Show that any monotone real function is Borel-measurable.

It can be shown that, writing c for the cardinality of the reals, there are c Borel subsets of $\mathbb{R}$, but 2^c Lebesgue-measurable subsets of $\mathbb{R}$. Since $A \subset \mathbb{R}$ belongs to $\mathcal{L}$ (resp. $\mathcal{B}$) iff $\mathbf{1}_A$ is Lebesgue (resp. Borel-)measurable, it follows that there are many Lebesgue-measurable functions that are not Borel-measurable. Nonetheless, it turns out that for any Lebesgue-measurable f there is a Borel-measurable function agreeing with f up to an m-null set.

3.3 Stability properties

Recall that we defined the Borel σ-field $\mathcal{B} = \mathcal{B}(\mathbb{R})$ by means of intervals of the form $(a, b]$. In exactly the same way we can use *rectangles* of the form $J = \prod_{i=1}^n (a_i, b_i]$ consisting of vectors $\mathbf{x} = (x_1, x_2, \ldots, x_n) \in \mathbb{R}^n$ with $a_i < x_i \leq b_i$ for all $i \leq n$, to define the Borel σ-field on $\mathbb{R}^n$: the collection $\mathcal{J}$ of such rectangles is a π-system, and we take $\mathcal{B}(\mathbb{R}^n) = \sigma(\mathcal{J})$. Clearly, this produces the same σ-field if we start with products of open sets, or even Borel sets, on the line.

Now fix a σ-finite measure space $(\Omega, \mathcal{F}, \mu)$. The following result provides a large class of 'new' measurable sets from 'old', again using the Monotone Class Theorem:

Theorem 3.6 *If $f_1, \ldots, f_n$ are measurable functions, $f_i : \Omega \to \mathbb{R}$ and if $G : \mathbb{R}^n \to \mathbb{R}$ is Borel-measurable (i.e. $G^{-1}(B) \in \mathcal{B}(\mathbb{R}^n)$ for all B in $\mathcal{B}$), then $g = G(f_1, \ldots, f_n) : \Omega \to \mathbb{R}$ is measurable.*

Proof Apply Exercise 3.1 to $\mathbf{f} = (f_1, \ldots, f_n) : \Omega \to \mathbb{R}^n$. If $\mathbf{B} \in \mathcal{B}(\mathbb{R}^n)$ is a 'rectangle' $\mathbf{B} = \prod_{i=1}^n B_i$ with $B_i \in \mathcal{B}$, then

$$
\begin{aligned}
\mathbf{f}^{-1}(\mathbf{B}) &= \{\omega \in \Omega : (f_1(\omega), \ldots, f_n(\omega)) \in \mathbf{B}\} \\
&= \{\omega \in \Omega : f_1(\omega) \in B_1, \ldots, f_n(\omega) \in B_n\} \\
&= \cap_{i=1}^n \{\omega \in \Omega : f_i(\omega) \in B_i\} \in \mathcal{F}.
\end{aligned}
$$

By Exercise 3.1, the class $\mathcal{C} = \{\mathbf{A} \in \mathcal{B}(\mathbb{R}^n) : \mathbf{f}^{-1}(\mathbf{A}) \in \mathcal{F}\}$ is a σ-field. It contains the π-system of rectangles. Hence it equals $\mathcal{B}(\mathbb{R}^n)$ by the Monotone Class Theorem. But then for any $B \in \mathcal{B}$ we have

$$
g^{-1}(B) = \{\omega \in \Omega : (f_1(\omega), \ldots, f_n(\omega)) \in G^{-1}(B)\},
$$

and $G^{-1}(B) \in \mathcal{B}(\mathbb{R}^n)$, so $g^{-1}(B) \in \mathcal{F}$.

Exercise 3.7 Suppose that f, g are measurable real-valued functions defined on $(\Omega, \mathcal{F})$. Show that:

(i) the set $\mathcal{L}^0(\mathcal{F})$ of measurable real-valued functions is a vector space and is closed under multiplication (i.e. $f.g$ is measurable);

(ii) for any a, the truncation of f given by $f^a(\omega) = a$ if $f(\omega) > a$, and $f^a(\omega) = f(\omega)$ if $f(\omega) \leq a$, is measurable;

(iii) f is measurable iff its positive part $f^+ = f.\mathbf{1}_{\{f>0\}}$ and its negative part $f^- = -f\mathbf{1}_{\{f\leq 0\}}$ are both measurable;

(iv) if f is measurable, then so is $|f| = f^+ + f^-$, but not conversely.

We formulate our next results for real-valued functions defined on an arbitrary measurable E subset of Ω: the restricted space $(E, \mathcal{F}_E, \mu_E)$ is of course again a measure space!

Passage to the limit does not destroy measurability – we just need the stability properties of $\mathcal{F}$.

Theorem 3.8 *If $\{f_n\}$ is a sequence of measurable functions defined on the set E in Ω, then the following are measurable functions also*

$$\max_{n\leq k} f_n, \ \min_{n\leq k} f_n, \ \sup_{n\in\mathbb{N}} f_n, \ \inf_{n\in\mathbb{N}} f_n,$$
$$\limsup_{n\to\infty} f_n, \ \liminf_{n\to\infty} f_n.$$

Proof (**Sketch**) It is sufficient to note that the following are measurable sets:

$\{\omega : (\max_{n\leq k} f_n)(\omega) > a\} = \bigcup_{n=1}^k \{\omega : f_n(\omega) > a\},$
$\{\omega : (\min_{n\leq k} f_n)(\omega) > a\} = \bigcap_{n=1}^k \{\omega : f_n(\omega) > a\},$
$\{\omega : (\sup_{n\geq k} f_n)(\omega) > a\} = \bigcup_{n=k}^\infty \{\omega : f_n(\omega) > a\},$
$\{\omega : (\inf_{n\geq k} f_n)(\omega) \geq a\} = \bigcap_{n=k}^\infty \{\omega : f_n(\omega) \geq a\}.$

The upper limit is $\limsup_{n\to\infty} f_n = \inf_{n\geq 1}\{\sup_{m\geq n} f_m\}$ and the above relations show that $h_n = \sup_{m\geq n} f_m$ is measurable; hence $\inf_{n\geq 1} h_n(x)$ is measurable. The lower limit is done similarly.

Corollary 3.9 *If a sequence f_n of measurable functions converges (pointwise), then the limit is also measurable.*

This is immediate since $\lim_{n\to\infty} f_n = \limsup_{n\to\infty} f_n$, which is measurable.

Remark 3.10 Note that for functions $f : \mathbb{R} \to \mathbb{R}$, Theorems 3.6 and 3.8 have counterparts for Borel functions, i.e. they remain valid upon replacing 'measurable' by 'Borel' throughout. However, things are slightly more complicated when we consider the role of null sets. On the one hand, changing a function on a null set cannot destroy its measurability, i.e. any measurable function which is altered on a null set remains measurable. However, as not all null sets are necessarily Borel sets, we cannot conclude similarly for Borel sets, and thus, in the case $\Omega = \mathbb{R}, E \in \mathcal{L}$, the following results have no natural 'Borel' counterparts.

Proposition 3.11 *If $f : E \to \mathbb{R}$ is measurable, $E \in \mathcal{F}$, $g : E \to \mathbb{R}$ is arbitrary, and the set $\{\omega : f(\omega) = g(\omega)\}$ is μ-null, then g is measurable.*

Proof Consider the difference $d(\omega) = g(\omega) - f(\omega)$. It is 0 except on a null set, so $\{\omega : d(\omega) > a\}$ is a null set if $a \geq 0$, and its complement is a null set if $a < 0$. Both sets are measurable and hence d is a measurable function. Thus $g = f + d$ is measurable.

Corollary 3.12 *If (f_n) is a sequence of measurable functions and $f_n(\omega) \to f(\omega)$ a.e. for ω in E, then f is measurable.*

Proof Let A be the μ-null set such that $f_n(\omega)$ converges for all $\omega \in C = E \setminus A$. Then $\mathbf{1}_C f_n$ converges everywhere to $g = \mathbf{1}_C f$, which is therefore measurable. But $f = g$ a.e., so f is also measurable.

Exercise 3.13 Let f_n be a sequence of measurable functions. Show that the set $E = \{\omega : f_n(\omega) \text{ converges}\}$ is measurable.

Since we are able to adjust a function f at will on a null set without altering its measurability properties, the following definition is a useful means of concentrating on the values of f that 'really matter' for integration theory, by identifying its bounds 'outside null sets':

Definition 3.14 Suppose $f : E \to \overline{\mathbb{R}}$ is measurable. The essential supremum $ess \sup f$ is $\inf\{z : f \leq z \ \mu\text{-a.e.}\}$ and the essential infimum ess inf is $\sup\{z : f \geq z \ \mu\text{-a.e.}\}$.

The following is then straightforward:

Exercise 3.15 (i) Show that if f, g are measurable functions, then $\operatorname{ess\,sup}(f + g) \leq \operatorname{ess\,sup} f + \operatorname{ess\,sup} g$.

(ii) Show that for measurable f, $\operatorname{ess\,sup} f \leq \sup f$. Show that these quantities coincide when f is continuous.

Although the class of measurable functions is very large, we can always approximate any *positive* measurable function 'from below' by simple functions. We call $\phi : \Omega \to \mathbb{R}$ *simple* if its range (i.e. the set of its values) is finite. As seen in Exercise 3.4, such functions *partition* Ω as the union of finitely many disjoint sets in $\mathcal{F}$; if ϕ takes the distinct values $\{a_1, \ldots, a_n\}$ and $A_i = \phi^{-1}(\{a_i\}) = \{\omega \in \Omega : \phi(\omega) = a_i\}$ for $i = 1, 2, \ldots, n$, then we can represent ϕ as $\phi = \sum_{i=1}^{n} a_i \mathbf{1}_{A_i}$.

The vector space $\mathcal{S}(\mathcal{F})$ of simple functions contains the basic building blocks from which other measurable functions can be built.

Proposition 3.16 *Suppose $f : \Omega \to [0, \infty)$ is measurable. Then the simple functions*

$$s_n = \sum_{k=1}^{n2^n} \frac{k-1}{2^n} \mathbf{1}_{A_k} + n \mathbf{1}_{B_n},$$

where $A_k = \left\{ \frac{k-1}{2^n} \leq f < \frac{k}{2^n} \right\}$ and $B_n = \{ f \geq n \}$ form an increasing sequence converging pointwise to f.

Figure 3.1 illustrates this proposition. Use it to provide a formal proof.

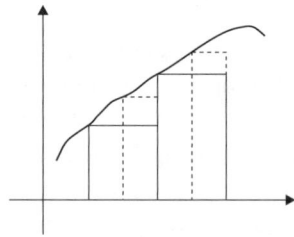

Fig. 3.1 Approximation by step functions

3.4 Random variables and independence

From now on let $(\Omega, \mathcal{F}, P)$ be a fixed probability space, and $X : \Omega \to \mathbb{R}$ a random variable. Recall that the σ-field *generated* by X is given by $\sigma(X) = \{X^{-1}(B) : B \in \mathcal{B}\}$.

In Exercise 3.4 we examined a number of special cases. This showed that what matters is 'how many distinct values' X takes, rather than what these values are. Note also that if $\mathcal{G}$ is any σ-field, then X is $\mathcal{G}$-measurable iff $\sigma(X) \subset \mathcal{G}$.

We can similarly define the σ-field generated by a sequence, or indeed any family $(X_\alpha)_{\alpha \in \Lambda}$, of random variables. Denote by $\sigma((X_\alpha)_{\alpha \in \Lambda})$ the smallest σ-field containing the family $\left(X_\alpha^{-1}(B) : \alpha \in \Lambda, B \in \mathcal{B} \right)$. A family of random variables indexed by some index set Λ is called a *stochastic process*. The main examples are when $\Lambda = \mathbb{N}$ or $\Lambda = [0, \infty)$. We usually think of the index set as representing *'time'*. In the first case we call the sequence $(X_n)_{n \geq 1}$ a *discrete-time process*, in the second, $(X_t)_{t \geq 0}$ is a *continuous-time process*. Stochastic processes are used to model the *dynamic behaviour* of random phenomena. We examine several key processes in later chapters.

Recall the definition of the *distribution* (or *law*) of a random variable $X : \Omega \to \mathbb{R}$: this is the probability measure P_X on $(\mathbb{R}, \mathcal{B})$ defined by $P_X(B) = P(X^{-1}(B)) = P(X \in B)$ for all $B \in \mathcal{B}$. In Exercise 2.10 we showed that this defines a probability measure on $\mathbb{R}$. We saw that P_X is determined by the (cumulative) *distribution function* $F_X : \mathbb{R} \to [0, 1]$ given by $F_X(x) = P_X((-\infty, x]) = P(X \leq x)$.

Its elementary properties were discussed in Exercise 2.12.

Definition 3.17 We say that two random variables X, Y are *identically distributed* if $F_X = F_Y$ (this is of course equivalent to $P_X = P_Y$). Write $X \overset{d}{=} Y$ for this.

Thus, in applications, we often do not specify the probability space on which X is defined: we prove results about particular distribution functions, and these then apply to any random variable with this distribution. X and Y need not have any values in common if $X \overset{d}{=} Y$; all that matters is that the probabilities of them being no greater than x should be equal for all real x. Clearly, however, if $X(\omega) = Y(\omega)$ for *almost all* $\omega \in \Omega$, i.e. if $\{\omega \in \Omega : X(\omega) \neq Y(\omega)\}$ is P-null, then $X \overset{d}{=} Y$.

We can extend the definitions to *random vectors*: for a finite number of random variables, let $\mathbf{X} = (X_1, X_2, \ldots, X_n)$ be a measurable function from Ω to $\mathbb{R}^n$, then the *distribution* of $\mathbf{X}$ (also called its *law*) is the probability $P_{\mathbf{X}}$ on $(\mathbb{R}^n, \mathcal{B}(\mathbb{R}^n))$ defined by $P_X(B) = P(\mathbf{X} \in B)$ for B in $\mathcal{B}(\mathbb{R}^n)$, and the *joint distribution* of $(X_1, \ldots, X_n)$ is then simply the distribution function $F_{\mathbf{X}} : \mathbb{R}^n \to [0, 1]$ given by

$$F_{\mathbf{X}}(x_1, \ldots, x_n) = P(X_1 \leq x_1, \ldots, X_n \leq x_n).$$

We come to the key concept of independence of random variables:

Definition 3.18 Random variables $X_1, X_2, \ldots, X_n$ are *independent* if for every choice of Borel sets $B_1, B_2, \ldots, B_n$

$$P(X_1 \in B_1, X_2 \in B_2, \ldots, X_n \in B_n) = \prod_{i=1}^{n} P(X_i \in B_i).$$

Any family of random variables is *independent* iff every finite subset is independent.

If the random variables are independent and all have the same distribution, then we call them an *independent and identically distributed (i.i.d.) family*. An example is given by the Bernoulli random variables making up our random walk in Section 1. Both this definition and independence of events (Definition 1.22) are special cases of the following:

Definition 3.19 Sub-σ-fields $(\mathcal{F}_k)_k$ of $\mathcal{F}$ are *independent* if for all choices of distinct indices $\{i_1, \ldots, i_n\}$ and F_{i_k} in $\mathcal{F}_{i_k}$ $(k \leq n)$ we have $P\left(\cap_{k=1}^{n} F_{i_k}\right) = \prod_{k=1}^{n} P(F_{i_k})$.

By a *sub-σ-field* we mean a σ-field $\mathcal{G} \subset \mathcal{F}$, which contains Ω. We now show that Definitions 3.19 and 1.22 are equivalent to the following (we go from the general to the particular here!):
(i) Random variables $(X_k)_{k \geq 1}$ are independent iff their generated σ-fields $(\sigma(X_k))_{k \geq 1}$ are independent.
(ii) Events $(F_k)_{k \geq 1}$ in $\mathcal{F}$ are independent iff the σ-fields $(\mathcal{F}_k)_{k \geq 1}$ are independent, where $\mathcal{F}_k = \{\emptyset, F_k F_k^c, \Omega\}$.

These definitions ensure that events $(F_k)_{k \geq 1}$ are independent iff their indicators $(\mathbf{1}_{F_k})_{k \geq 1}$ are independent. To see that they are equivalent to the earlier versions, we again use π-systems to simplify matters – we can restrict attention to the case $n = 2$, as induction would do the rest:

Lemma 3.20 *Suppose the sub-σ-fields $\mathcal{F}_1, \mathcal{F}_2$ in $\mathcal{F}$ are generated by π-systems $\mathcal{C}_1, \mathcal{C}_2$. Then $\mathcal{F}_1$ and $\mathcal{F}_2$ are independent iff*

$$P(C_1 \cap C_2) = P(C_1)P(C_2)$$

for all choices of C_1 in $\mathcal{C}_1$ and C_2 in $\mathcal{C}_2$.

Proof If $P(C_1 \cap C_2) = P(C_1)P(C_2)$ for fixed C_1 in $\mathcal{C}_1$, the non-negative finite set functions $C_2 \mapsto P(C_1 \cap C_2)$ and $C_2 \to P(C_1)P(C_2)$, defined on $(\Omega, \mathcal{F}_2)$, are clearly countably additive and both equal $P(C_1)$ if $C_2 = \Omega$. Thus they are measures that agree on the π-system $\mathcal{C}_2$, hence by Theorem 2.9 they agree on $\sigma(\mathcal{C}_2) = \mathcal{F}_2$. This means that $P(C_1 \cap F_2) = P(C_1)P(F_2)$ for all F_2 in $\mathcal{F}_2$. Repeating the above argument with fixed F_2, we see that the measures agree on $\sigma(\mathcal{C}_1) = \mathcal{F}_1$, so that $P(F_1 \cap F_2) = P(F_1)P(F_2)$ for all $F_1 \in \mathcal{F}_1, F_2 \in \mathcal{F}_2$, hence the σ-fields are independent.

For any random variable X, $\pi(X) = (\{\omega \in \Omega : X(\omega) \leq x\})_{x \in \mathbb{R}}$ is a π-system, so if X, Y are independent in the above sense, then $\pi(X)$ and $\pi(Y)$ are independent, hence so are $\sigma(X)$ and $\sigma(Y)$. This proves (i), and (ii) follows at once. Induction on n provides the proof for the general case.

The main result we need is that random variables are independent iff their joint distribution function is the product of their individual distribution functions – in other words, we need only check Definition 3.18 for intervals $B_i = (-\infty, x_i), i \leq n$.

Theorem 3.21 *$X_1, X_2, \ldots, X_n$ are independent iff, for the random vector $\mathbf{X} = (X_1, X_2, \ldots, X_n)$, the joint distribution function satisfies*

$$F_{\mathbf{X}}(x_1, \ldots, x_n) = \prod_{i=1}^{n} F_{X_i}(x_i) \text{ for all } (x_1, \ldots, x_n) \in \mathbb{R}^n.$$

Proof The necessity is clear, as intervals are Borel sets. For the sufficiency, let $\mathcal{C}_1$ be the family of Borel sets $B_1 \in \mathcal{B}$ such that for all real $x_2, x_3, \ldots, x_n$ we have

$$P(X_1 \in B_1, X_i \leq x_i, i = 2, \ldots, n) = P(X_1 \in B_1) \prod_{i=2}^{n} F_{X_i}(x_i).$$

Then $\mathcal{C}_1$ contains every interval of the form $(-\infty, x]$ so it contains the π-system that generates $\mathcal{B}$. If we can show that $\mathcal{C}_1$ is a d-system, the Monotone Class Theorem implies that our identity holds for all Borel sets B_1 on the line.

Assume for a moment that this has been done. We can then iterate the procedure for each index: for example, we let $\mathcal{C}_2$ be the class of Borel sets B_2 for which

$$P(X_1 \in B_1, X_2 \in B_2, X_i \leq x_i, i = 3, \ldots, n)$$
$$= P(X_1 \in B_1)P(X_2 \in B_2) \prod_{i=3}^{n} F_{X_i}(x_i)$$

for all $B_1 \in \mathcal{B}$ and choices $x_3, \ldots, x_n$ in $\mathbb{R}$. This class again contains all intervals of the form $(-\infty, x]$ and will be a d-system just like $\mathcal{C}_1$. This will complete the proof after n steps.

So it remains to show that $\mathcal{C}_1$ is a d-system. This is entirely routine and is left as an exercise.

Examples of the basic *discrete* probability distributions were given earlier. At the other extreme are the *absolutely continuous* distributions: then

$$P(X \leq x) = F_X(x) = \int_{-\infty}^{x} f_X(t)dt$$

for some positive function f_X (the *density* of X).

In the examples below, f_X is piecewise continuous, therefore the Riemann-integral is well-defined – in general we use the Lebesgue integral described in Chapter 1 and defined formally in the next chapter.

Example 3.22 Let X be a random variable.

(i) X is *uniformly distributed* on $[a, b]$ if it has the constant density $f_X(x) = \frac{1}{b-a} \mathbf{1}_{[a,b]}(x)$.

(ii) X is *normally distributed* with parameters $\mu \in \mathbb{R}, \sigma^2 > 0$ when $f_X(x) = \frac{1}{\sqrt{2\pi\sigma^2}} \exp\left(-\frac{1}{2}\left(\frac{x-\mu}{\sigma}\right)^2\right)$. We say $X \sim \mathcal{N}(\mu, \sigma^2)$. It is easily checked that f_X is a density ($f_X \geq 0$, $\int_{-\infty}^{\infty} f_X(t)dt = 1$) and that $\frac{X-\mu}{\sigma} \sim \mathcal{N}(0, 1)$ (*standard normal*) iff $X \sim \mathcal{N}(\mu, \sigma^2)$.

(iii) $X \geq 0$ is *exponentially distributed* with parameter $\lambda > 0$ if $f_X(x) = \lambda e^{-\lambda x} 1_{[0,\infty)}(x)$. Just as for geometric distributions, this describes a 'waiting time', in this case the interarrival times of a Poisson process.

The next example is analogous to the negative binomial distribution.

(iv) $X \geq 0$ is *gamma distributed* with parameters $\alpha > 0, \lambda > 0$ if $f_X(x) = \frac{1}{\Gamma(\alpha)} \lambda^\alpha e^{-\lambda x} x^{\alpha-1} 1_{[0,\infty)}(x)$. Here $\Gamma(\alpha) = \int_0^\infty y^{\alpha-1} e^{-y} dy$ is the gamma function. Note that $\Gamma(1) = 1$, so (iii) is the special case $\alpha = 1$. The *chi-squared* distributions ($X \sim \chi^2(k)$) are the special case $\lambda = \frac{1}{2}, \alpha = \frac{k}{2}$.

The sum of k independent exponentially distributed random variables with the parameter λ is gamma-distributed, and the sum of the squares of k independent standard normal random variables is $\chi^2(k)$.

4

Integration and expectation

The abstractly defined *integral* $\int_\Omega f \, d\mu$ of a function $f : \Omega \to \mathbb{R}$, relative to a measure μ defined on a given measurable space $(\Omega, \mathcal{F})$, follows a pattern first established for the Lebesgue measure m on $(\mathbb{R}, \mathcal{B})$. We begin with indicator functions, extend to simple functions by linearity, and then by limits to arbitrary positive measurable functions. Since we will allow the value $+\infty$ for the integral, however, we need a little care when extending further to real-valued measurable functions. The natural extension is to use Exercise 3.7 (iii): writing $f = f^+ - f^-$ we know that f is measurable iff both the positive functions on the right are measurable. Having defined their integrals as elements of $[0, \infty]$, we must avoid the expression $\infty - \infty$ in defining the integral of f. Thus we restrict attention to functions f where $\int_\Omega |f| \, d\mu < \infty$. We call this space $\mathcal{L}^1(\Omega, \mathcal{F}, \mu)$; when the context is clear, we shorten this to $\mathcal{L}^1(\Omega)$, $\mathcal{L}^1(\mathcal{F})$ or $\mathcal{L}^1(\mu)$ (depending on which aspect we wish to emphasise), or even just to $\mathcal{L}^1$. We will show that $\mathcal{L}^1$ is a vector space and that the map $f \to \int_\Omega f \, d\mu$ is *linear.* Thus we shall also write $\mu(f) = \int_\Omega f \, d\mu$. For a probability space $(\Omega, \mathcal{F}, P)$ the integral of the random variable $X : \Omega \to \mathbb{R}$ is called the *expectation* of X and written $\mathbb{E}[X] = \int_\Omega X \, dP$. Note that sums are a special case of integrals: $\mu(\{n\}) = 1$ for each n defines a (σ-finite!) measure on $(\mathbb{N}, 2^{\mathbb{N}})$ for real $(a_n)_{n \geq 1}$ with $\sum_{n=1}^{\infty} |a_n| < \infty$. In that case the map $f : n \to a_n$ is μ-integrable (see below), and we set $\int_{\mathbb{N}} f \, d\mu = \sum_{n=1}^{\infty} a_n$.

4.1 Integrals of positive measurable functions

Definition 4.1 Fix a σ-finite measure space $(\Omega, \mathcal{F}, \mu)$. For $A \in \mathcal{F}$, define the μ-integral of its indicator $\mathbf{1}_A$ by

$$\mu(\mathbf{1}_A) = \int_\Omega \mathbf{1}_A d\mu = \mu(A).$$

For a *positive simple* function, $\phi = \sum_{i=1}^n a_i \mathbf{1}_{A_i}$, where $a_i > 0$ and $(A_i)_{i \le n} \in \mathcal{F}$ are pairwise disjoint; we extend by linearity

$$\mu(\phi) = \int_\Omega \phi d\mu = \sum_{i=1}^n a_i \mu(A_i).$$

Since we allow infinite values, we need to extend arithmetic to $[0, \infty]$. For this, we need only add, for $a \in [0, \infty]$

$$a + \infty = \infty + a = \infty,$$
$$a.\infty = \infty = \infty.a \text{ if } a > 0, 0.\infty = 0 = \infty.0$$

to normal arithmetic in order to to remain consistent.

Exercise 4.2 Check that $\mu(\phi)$ is well-defined, i.e. independent of the representation of ϕ.

Example 4.3 For Lebesgue measure m and a real function f, we shall write the integral as $\int_{-\infty}^\infty f(x)m(dx)$. For example:
(i) $\int_\mathbb{R} \mathbf{1}_\mathbb{Q} dm = 1.m(\mathbb{Q}) + 0.m(\mathbb{R} \backslash \mathbb{Q}) = 0$.
(ii) If $[x]$ is the integer part of x, then $\int_0^2 [x^2] m(dx) = 5 - \sqrt{3} - \sqrt{2}$. ($[x^2]$ changes at $1, \sqrt{2}, \sqrt{3}, 2!$).

We write $\mathcal{S}^+(\mathcal{F})$ for the cone of positive simple functions and $\mathcal{M}(\mathcal{F})$ (resp. $\mathcal{M}^+(\mathcal{F})$) for the vector space (resp. cone) of $\mathcal{F}$-measurable (resp. positive $\mathcal{F}$-measurable) functions. Clearly, $\mathcal{S}^+(\mathcal{F}) \subset \mathcal{M}^+(\mathcal{F})$.

It is helpful to allow integrals to be taken over arbitrary sets $E \in \mathcal{F}$: simply define $\int_E f d\mu$ as $\mu(f \mathbf{1}_E) = \int_\Omega f \mathbf{1}_E d\mu$. The following basic facts are now easy to check for $\mathcal{S}^+(\mathcal{F})$.

Exercise 4.4 Let $\phi, \psi \in \mathcal{S}^+(\mathcal{F})$ and $E, F \in \mathcal{F}$ be given. Show:
(i) $\mathcal{S}^+(\mathcal{F})$ is a cone and a lattice, i.e. $\phi + \psi$ and $c\phi$ are in $\mathcal{S}^+(\mathcal{F})$ for
$c \geq 0$ and $\phi \vee \psi = \max(\phi, \psi)$, $\phi \wedge \psi = \min(\phi, \psi)$ are in $\mathcal{S}^+(\mathcal{F})$.
(ii) If $\phi = \psi$ a.e.(μ), then $\int_E \phi d\mu = \int_E \psi d\mu$.
(iii) The map $\phi \rightarrow \int_E \phi d\mu$ is positive-linear, i.e.

$$\int_E (\phi + \psi) d\mu = \int_E \phi d\mu + \int_E \psi d\mu,$$

$$\int_E (c\phi) d\mu = c \int_E \phi d\mu \text{ for } c > 0.$$

(iv) The map $\phi \rightarrow \int_E \phi d\mu$ is monotone, i.e.

$$\text{if} \phi \leq \psi \text{ a.e.}(\mu), \text{ then } \int_E \phi d\mu \leq \int_E \psi d\mu.$$

(v) If $E \cap F = \emptyset$, then $\int_{E \cup F} \phi d\mu = \int_E \phi d\mu + \int_F \phi d\mu$.

We extend the integral to $f \in \mathcal{M}^+(\mathcal{F})$.

Definition 4.5 Let

$$Y(f) = \left\{ \int_\Omega \phi d\mu : \phi \in \mathcal{S}(\mathcal{F}) : 0 \leq \phi \leq f \right\}$$

and define

$$\mu(f) = \int_\Omega f d\mu = \sup Y(f).$$

Exercise 4.6 For $\phi \in \mathcal{S}^+(\mathcal{F})$, we now have two definitions of the integral $\int_\Omega \phi d\mu$, since $\mathcal{S}^+(\mathcal{F}) \subset \mathcal{M}^+(\mathcal{F})$. Show that the two definitions coincide.

For $E \in \mathcal{F}$, we write $\int_E f d\mu = \mu(f \mathbf{1}_E)$. Clearly, we could equally well define $\int_E f d\mu = \mu(f \mathbf{1}_E)$ as $\sup Y(E, f)$, where

$$Y(E, f) = \left\{ \int_E \phi d\mu : \phi \in \mathcal{S}(\mathcal{F}), 0 \leq \phi \leq f \text{ on } E \right\}.$$

Elementary properties of the integral, proved in Exercise 4.4 for simple functions, extend easily to $\mathcal{M}^+(\mathcal{F})$:

Proposition 4.7 *Let $f, g \in \mathcal{M}^+(\mathcal{F})$ and $E, F \in \mathcal{F}$ be given. Then:*
(i) $f \leq g$ on E implies $\int_E f d\mu \leq \int_E g d\mu$.

(ii) $F \subset E$ *implies* $\int_F f d\mu \leq \int_E f d\mu$.
(iii) $a > 0$ *implies* $\int_E (af) d\mu = a \int_E f d\mu$.
(iv) $\mu(E) = 0$ *implies* $\int_E f d\mu = 0$, *and* $\int_E f d\mu = 0$ *implies* $f = 0$
a.e. (μ) *on* E.
(v) $E \cap F = \emptyset$ *implies* $\int_{E \cup F} f d\mu = \int_E f d\mu + \int_F f d\mu$.

Proof (i) and (iii) are obvious. For (ii), note that if $\phi \in \mathcal{S}^+(\mathcal{F})$ and $\phi \leq f$ on F, we can extend it to a simple function $\psi \leq f$ on E by setting $\psi = 0$ on $E \backslash F \in \mathcal{F}$. Clearly, $\int_F \phi d\mu = \int_E \psi d\mu$.

To prove (iv), let $\phi = \sum_{i=1}^{n} a_i \mathbf{1}_{A_i} \in \mathcal{S}^+(\mathcal{F})$ and $\phi \leq f$ on E. Then

$$\int_E \phi d\mu = \sum_{i=1}^{n} a_i \mu(E \cap A_i) = 0,$$

so that $Y(E, f) = \{0\}$. Hence $\int_E f d\mu = 0$.

On the other hand, restricting to E, let $A = \{f > 0\}$ and $A_n = \{f > \frac{1}{n}\}$ for each $n \geq 1$. Then $A_n \uparrow A$. If $\mu(A) > 0$, we can find n with $\mu(A_n) > 0$. Now $\phi = \frac{1}{n} \mathbf{1}_{A_n} \leq f$, and $\phi \in \mathcal{S}^+(\mathcal{F})$, so that

$$0 = \int_E f d\mu \geq \int_E \phi d\mu = \frac{1}{n} \mu(A_n) > 0,$$

a contradiction. Hence $\int_E f d\mu = 0$ implies $f = 0$ a.e. (μ).

To prove (v), note that by Exercise 4.4 we obtain $Y(E \cup F, f) = Y(E, f) + Y(F, f)$ when E and F are disjoint. Taking suprema, we have

$$\int_{E \cup F} f d\mu \leq \int_E f d\mu + \int_F f d\mu.$$

For the opposite inequality, take simple functions $\phi \leq f$ on E and $\psi \leq f$ on F to construct a simple function $\theta = \phi \mathbf{1}_E + \psi \mathbf{1}_F \leq f$ on $E \cup F$, so that

$$\int_E \phi d\mu + \int_F \psi d\mu = \int_{E \cup F} \theta d\mu \leq \int_{E \cup F} f d\mu.$$

The RHS is an upper bound independent of the choices of ϕ, ψ. Fix ψ and take the sup over $Y(E, f)$ then take the sup over $Y(F, f)$ to conclude that

$$\int_{E \cup F} f d\mu \geq \int_E f d\mu + \int_F f d\mu.$$

Example 4.8 On $[0, 1]$ with Lebesgue measure m, define the function f by setting $f = 0$ on the Cantor set C and for each $k \geq 1$, let $f(x) = k$ on each of the intervals of length 3^{-k} removed from $[0, 1]$ in the construction of C. To find $\int_0^1 f(x)m(dx)$, we write $g = \sum_{k=1}^{\infty} k\mathbf{1}_{A_k}$, where A_k is the union of the 2^{k-1} disjoint intervals removed at the kth stage. We have $g = f$ a.e.(m), so their integrals are equal. To find $\int_0^1 g(x)m(dx)$, note that for each $n \geq 1$, $\phi_n = \sum_{k=1}^{n} k\mathbf{1}_{A_k} \in \mathcal{S}^+(\mathcal{F})$ and $0 \leq \phi_n \leq g$. The integrals $\int_0^1 \phi_n(x)m(dx) = \sum_{k=1}^{n} km(A_k)$ have $\sum_{k=1}^{\infty} km(A_k)$ as their supremum, and $\phi_n \uparrow g$ pointwise; hence

$$\int_0^1 g(x)m(dx) = \sum_{k=1}^{\infty} km(A_k) = \sum_{k=1}^{\infty} k2^{k-1}3^{-k}$$
$$= \frac{1}{3}\sum_{k=1}^{\infty} k\left(\frac{2}{3}\right)^{k-1} = 3.$$

The key property of the integral is its stability under limits, as Example 4.8 suggests. The first convergence theorems belong in the space $\mathcal{M}^+(\mathcal{F})$.

Theorem 4.9 *Monotone Convergence Theorem (MCT)*
If $(f_n)_{n \geq 1}$ in $\mathcal{M}^+(\mathcal{F})$ and $f_n \uparrow f$ on $E \in \mathcal{F}$, then $\int_E f_n d\mu \uparrow \int_E f d\mu$.

Proof Note that $\int_E f d\mu$ is well-defined, since $f = \lim_n f_n$ is in $\mathcal{M}^+(\mathcal{F})$.

We first prove the theorem for $E = \Omega$. By Proposition 4.7, the sequence $(\int_\Omega f_n d\mu)_{n \geq 1}$ increases to $L = \lim_n \int_\Omega f_n d\mu \leq \int_\Omega f d\mu$. We need to show that $\int_\Omega f d\mu \leq L$. Take $c \in (0, 1)$, choose a simple function $\phi = \sum_{i=1}^{m} a_i \mathbf{1}_{E_i} \leq f$, and let $A_n = \{f_n \geq c\phi\}$. The $(A_n)_n$ increase with n and have union Ω. For each $n \geq 1$

$$\int_\Omega f_n d\mu \geq \int_{A_n} f_n d\mu \geq c\int_{A_n} \phi d\mu = c\sum_{i=1}^{m} a_i \mu(A_n \cap E_i).$$

As $n \to \infty$, the sum on the right converges to $c \sum_{i=1}^{m} a_i \mu(E_i) = c \int_{\Omega} \phi d\mu$. So $L \geq c \int_{\Omega} \phi d\mu$. But $c < 1$ was arbitrary, so $L \geq \int_{\Omega} \phi d\mu$, and as $\phi \leq f$ was arbitrary, it follows that $L \geq \int_{\Omega} f d\mu$. For general $E \in \mathcal{F}$, the result follows on applying the above to $g_n = f_n \mathbf{1}_E$, $g = f \mathbf{1}_E$ instead.

Apply this to the partial sums of $\sum_{n \geq 1} g_n$ in $\mathcal{M}^+(\mathcal{F})$:

Corollary 4.10 *If $g_n \in \mathcal{M}^+(\mathcal{F})$ and $\sum_{n \geq 1} g_n(\omega)$ converges a.e.(μ) on E, then $\int_E \left(\sum_{n=1}^{\infty} g_n \right) d\mu = \sum_{n=1}^{\infty} \int_E g_n d\mu$.*

Example 4.11 The MCT fails without the requirement that (f_n) is increasing. For an example on $\mathbb{R}$, let $f_n(x) = n\mathbf{1}_{(0, \frac{1}{n})}(x)$. Clearly, $f_n(x) \to 0$ for all x, but $\int f_n(x) \, dx = 1$.

Our next Corollary generalises what we proved for indicator functions in Proposition 1.14 and Exercise 1.15:

Corollary 4.12 *(The Fatou Lemmas)*
For any sequence (f_n) in $\mathcal{M}^+(\mathcal{F})$ and $E \in \mathcal{F}$:
(i) $\int_E (\liminf_n f_n) d\mu \leq \liminf_n \int_E f_n d\mu$.
(ii) If there is $g \in \mathcal{M}^+(\mathcal{F})$ with finite integral and $f_n \leq g$ for all n, then $\lim_n \sup_n \int_E f_n d\mu \leq \int_E (\limsup_n f_n) d\mu$.

Proof (i) Let $g_k = \inf_{n \geq k} f_n$, then

$$g_k \uparrow \sup_{k \geq 1} \inf_{n \geq k} f_n = \liminf_n f_n,$$

and for $n \geq k$, $f_n \geq g_k$. So $\int_E f_n d\mu \geq \int_E g_k d\mu$ for every $n \geq k$.

Hence $\int_E g_k d\mu \leq \inf_{n \geq k} \int_E f_n d\mu$ for each $k \geq 1$. Now apply MCT to (g_k), so that

$$\int_E (\liminf_n f_n) d\mu = \lim_k \int_E g_k d\mu$$

$$\leq \sup_{k \geq 1} \inf_{n \geq k} \int_E f_n d\mu = \liminf_n \int_E f_n d\mu.$$

(ii) Since $h_n = g - f_n$ is in $\mathcal{M}^+(\mathcal{F})$ for each n, and the integrals are finite, we can apply (i) to $(h_n)_n$

$$\int_E g\,d\mu - \limsup_n \int f_n\,d\mu = \liminf_n \int_E h_n\,d\mu$$
$$\geq \int_E (\liminf_n h_n)d\mu = \int_E g\,d\mu - \int_E (\limsup_n f_n)d\mu.$$

Example 4.13 It is easy to find examples where these inequalities are strict: on $\mathbb{R}$ let $f_n = \mathbf{1}_{(n,n+1]}$, so that $f_n \to 0$ pointwise, but $\int_\mathbb{R} f_n\,dm = 1$ for every n.

Remark 4.14 We observed that when $\mu = P$ and $f_n = \mathbf{1}_{A_n}$, then, since $\liminf_n \mathbf{1}_{A_n} = \mathbf{1}_{\liminf_n A_n}$, the Fatou Lemmas reduce to inequalities we found in the proof of Proposition 1.14 and in Exercise 1.15. Note similarly that, when $A_n \uparrow A$, the MCT applied to their indicators says that $P(A_n) \uparrow P(A)$, as was shown in Lemma 1.12(ii). Finally, Corollary 4.10 then reduces to a re-statement of the countable additivity of P.

Exercise 4.15 Verify that the integral is positive-linear on $\mathcal{M}^+(\mathcal{F})$: when $a, b \geq 0$ and $f, g \in \mathcal{M}^+(\mathcal{F})$ then for E in $\mathcal{F}$

$$\int_E (af + bg)d\mu = a\int_E f\,d\mu + b\int_E g\,d\mu.$$

(*Hint*: use the MCT on sequences in $\mathcal{S}^+(\mathcal{F})$ that approximate f, g from below, having applied Exercise 4.4 (iii).)

4.2 The vector space $\mathcal{L}^1$ of integrable functions

To define the integral for general real-valued functions, we simply employ the identity $f = f^+ - f^-$, recalling that $f \in \mathcal{M}(\mathcal{F})$ iff both its positive part f^+ and negative part f^- belong to $\mathcal{M}^+(\mathcal{F})$. Thus also $|f| \in \mathcal{M}^+(\mathcal{F})$, so that $\int_\Omega |f|\,d\mu$ is well-defined. We now require it to be *finite*, in order to avoid situations where both right-hand terms in the definition below are $+\infty$:

Definition 4.16 The function $f \in \mathcal{M}(\mathcal{F})$ is *μ-integrable over E in $\mathcal{F}$* if $\int_E |f|\, d\mu < \infty$ and its *(μ)-integral over E* is defined as

$$\int_E f d\mu = \int_E f^+ d\mu - \int_E f^- d\mu.$$

Denote the set of μ-integrable functions $E \to \mathbb{R}$ by $\mathcal{L}^1(E, \mathcal{F}_E, \mu_E)$ or simply by $\mathcal{L}^1(\mu)$.

Since $f \le |f|$ and $-f \le |f|$, we see that

$$\left| \int_E f d\mu \right| \le \int_E |f|\, d\mu.$$

Theorem 4.17 $\mathcal{L}^1(\mu)$ *is a vector space and the map $f \to \int_E f d\mu$ is linear on $\mathcal{L}^1$.*

Proof If $a, b \in \mathbb{R}$ and $f, g \in \mathcal{L}^1$, then $|af + bg| \le |a|\,|f| + |b|\,|g|$, so that $\int_E |af + bg|\, d\mu$ is finite. Hence $\mathcal{L}^1$ is a vector space.

To prove the linearity of the integral, note that by Exercise 4.15 the integral is positive-linear on $\mathcal{M}^+(\mathcal{F})$, and consider the difference of h_1, h_2 in $\mathcal{M}^+(\mathcal{F})$: we have

$$h_1 + (h_1 - h_2)^- = (h_1 - h_2)^+ + h_2$$

and both sides are in $\mathcal{M}^+(\mathcal{F})$, so that we can apply the additivity of the integral on $\mathcal{M}^+(\mathcal{F})$ to conclude that

$$\int_E h_1 d\mu + \int_E (h_1 - h_2)^- d\mu = \int_E (h_1 - h_2)^+ d\mu + \int_E h_2 d\mu.$$

Rearrange to obtain

$$\int_E (h_1 - h_2) d\mu = \int_E h_1 d\mu - \int_E h_2 d\mu.$$

Now for $f, g \in \mathcal{L}^1$ we set $h_1 = f^+ + g^+$, $h_2 = f^- + g^-$, so that $h_1 - h_2 = f + g$ and thus that

$$\int_E (f + g) d\mu = \int_E h_1 d\mu - \int_E h_2 d\mu = \int_E f d\mu + \int_E g d\mu.$$

Finally, Exercise 4.15 shows that $a \int_E f d\mu = \int_E (af) d\mu$ for $a > 0$ if $f \in \mathcal{L}^1$. For $a < 0$, we have

$$\int_E (af) d\mu = \int_E (af)^+ d\mu - \int_E (af)^- d\mu$$

$$= \int_E (-a) f^- d\mu - \int_E (-a) f^+ d\mu = a \int_E f d\mu.$$

The integral determines the integrand almost everywhere:

Theorem 4.18 *Let f, g in $\mathcal{L}^1(\Omega, \mathcal{F}, \mu)$. If $\int_A f d\mu \leq \int_A g d\mu$ for all $A \in \mathcal{F}$, then $f \leq g$ a.e. (μ). In particular, if $\int_A f d\mu = \int_A g d\mu$ for all $A \in \mathcal{F}$, then $f = g$ a.e. (μ).*

Proof By additivity of the integral, it is sufficient to show that $\int_A h \, d\mu \geq 0$ for all $A \in \mathcal{F}$ implies $h \geq 0$ and then take $h = g - f$. Write $A = \{h < 0\}$; then $A = \bigcup A_n$, where $A_n = \{h \leq -\frac{1}{n}\}$. By monotonicity of the integral

$$0 \leq \int_{A_n} h \, d\mu \leq \int_{A_n} \left(-\frac{1}{n}\right) d\mu = -\frac{1}{n} \mu(A_n).$$

This can only happen if $\mu(A_n) = 0$. The sequence of sets A_n increases with n, hence $\mu(A) = 0$, and so $h \geq 0$ a.e. (μ).

A similar argument shows that if $\int_A h \, d\mu \leq 0$ for all A, then $h \leq 0$ a.e. (μ). This implies the second claim of the theorem: if $h = g - f$ and $\int_A h \, d\mu$ is both non-negative and non-positive, then $h \geq 0$ and $h \leq 0$ a.e., thus $h = 0$ a.e.

We can construct many interesting measures, especially probability distributions, with the next simple result.

Proposition 4.19 $A \mapsto \int_A f \, d\mu$ *is a measure if $f \geq 0$ is in $\mathcal{L}^1$.*

Proof Let $\nu(A) = \int_A f \, d\mu$. To prove $\nu(\bigcup_i E_i) = \sum_i \nu(E_i)$ for pairwise disjoint E_i, consider the sequence $g_n = f \mathbf{1}_{\bigcup_{i=1}^n E_i}$ and note that (g_n) increases to g, where $g = f \mathbf{1}_{\bigcup_{i=1}^\infty E_i}$. Now $\int g \, d\mu = \nu\left(\bigcup_{i=1}^\infty E_i\right)$, and

$$\int g_n \, d\nu = \int_{\bigcup_{i=1}^n E_i} f d\mu = \sum_{i=1}^n \int_{E_i} f d\mu = \sum_{i=1}^n \nu(E_i)$$

so the Monotone Convergence Theorem (MCT) completes the proof.

The limit theorem which turns out to be the most useful in practice states that the integrals converge for an a.e. convergent sequence which is *dominated* by an integrable function. Fatou's Lemma holds the key to the proof.

Theorem 4.20 *Dominated Convergence Theorem (DCT)*
Suppose $E \in \mathcal{F}$. Let (f_n) be a sequence of measurable functions such that $|f_n| \le g$ a.e.(μ) on E for all $n \ge 1$, where g is integrable over E. *If $f = \lim_{n \to \infty} f_n$ a.e., then f is integrable over E and $\lim_{n \to \infty} \int_E |f_n - f| \, d\mu = 0$.*
In particular, $\lim_{n \to \infty} \int_E f_n(x) \, d\mu = \int_E f \, d\mu$.

Proof $|f_n - f| \le 2g$ and $\int_E 2g d\mu < \infty$, so the second Fatou Lemma yields

$$\limsup_{n \to \infty} \int_E |f_n - f| \, d\mu \le \int_E \limsup_{n \to \infty} |f_n - f| \, d\mu = 0.$$

This proves the first claim. The second follows since

$$\left| \int_E f_n d\mu - \int_E f d\mu \right| = \left| \int_E (f_n - f) d\mu \right| \le \int_E |f_n - f| \, d\mu.$$

Example 4.21 Example 4.11 also applies here: for $f_n = n1_{[0, \frac{1}{n}]}$, no integrable g can dominate f_n. The least upper bound is $g(x) = \sup_n f_n(x)$, $g(x) = k$ on $\left(\frac{1}{k+1}, \frac{1}{k} \right]$, so

$$\int g(x) \, dm = \sum_{k=1}^{\infty} k \left(\frac{1}{k} - \frac{1}{k+1} \right) = \sum_{k=1}^{\infty} \frac{1}{k+1} = +\infty.$$

A typical use of the DCT yields the following:

Exercise 4.22 If $f : \mathbb{R} \to \mathbb{R}$ is integrable, define $g_n = f1_{[-n,n]}$ and $h_n = \min(f, n)$ (both truncate f in some way: the g_n vanish outside a bounded interval, the h_n are bounded). Show that $\int |f - g_n| \, dm \to 0$, $\int |f - h_n| \, dm \to 0$.

Another very useful consequence of the DCT allows us to integrate infinite sums:

Theorem 4.23 *(Beppo–Levi) Suppose $\sum_{k=1}^{\infty} \int |f_k| d\mu \leq \infty$ Then the series $\sum_{k=1}^{\infty} f_k(\omega)$ converges for almost all ω, its sum is integrable, and $\int \sum_{k=1}^{\infty} f_k d\mu = \sum_{k=1}^{\infty} \int f_k d\mu$.*

Proof   Apply Corollary 4.10 with $g_k = |f_k|$ to conclude that if $\phi(\omega) = \sum_{k=1}^{\infty} |f_k(\omega)|$, then $\int_E \phi d\mu = \sum_{k=1}^{\infty} \int_E |f_k| d\mu$ is finite. Thus ϕ is a.e.(μ) finite-valued, and so the series $\sum_{k\geq 1} f_k$ converges absolutely (and therefore converges) a.e.(μ). Denote the sum by $f(\omega)$ wherever the series converges, and put $f(\omega) = 0$ otherwise.

The partial sums $\left(\sum_{k=1}^{n} f_k\right)_n$ are bounded by the integrable function ϕ, so the DCT applies and we have

$$\int_E f d\mu = \lim_n \int_E \left(\sum_{k=1}^{n} f_k\right) d\mu = \lim_n \sum_{k=1}^{n} \int_E f_k d\mu,$$

which proves the theorem.

4.3 Riemann v. Lebesgue integrals

To calculate the Lebesgue integral $\int_{[a,b]} f dm$ of the real function $f : [a,b] \to \mathbb{R}$ we must relate it to the *Riemann integral* $\int_a^b f(x) dx$ familiar from calculus. There are many Lebegue-integrable functions – i.e. elements of the vector space $\mathcal{L}^1(\mathbb{R}, \mathcal{L}, m)$ – that are not Riemann-integrable. To begin with, Riemann integrals are computed over intervals and deal only with bounded functions (for unbounded functions we seek 'improper' integrals by means of a second limit procedure). Lebesgue integrals can be found over any Borel set and can handle many unbounded functions. But even for bounded functions on intervals, the Lebesgue integral has wider scope. As we show below, a bounded function $f : [a,b] \to \mathbb{R}$ is Riemann-integrable iff it is a.e.(m) continuous. However, $f = \mathbf{1}_{\mathbb{Q} \cap [0,1]}$, is nowhere continuous, but is clearly in $\mathcal{L}^1([0,1])$, with integral $m(\mathbb{Q}) = 0$.

Nonetheless, we need the 'recipe' provided by the Fundamental Theorem of Calculus to calculate actual integrals for many real functions. To use it, we need to know that the Lebesgue and Riemann integrals of a function f coincide whenever they both exist. But this follows from the fact that for any partition $\mathcal{P}$ of $[a, b]$ the upper ($U_\mathcal{P}$) and lower ($L_\mathcal{P}$) Riemann sums are just the integrals of simple functions $u_\mathcal{P}$ resp. $l_\mathcal{P}$: let $\mathcal{P}$ divide $[a, b]$ into n subintervals $[a_{i-1}, a_i]$, write $\Delta_i = a_i - a_{i-1}$ and $M_i = \sup_{a_{i-1} \leq x \leq a_i} f(x)$, then $u_\mathcal{P} = \sum_{i=1}^{n} M_i \mathbf{1}_{[a_{i-1}, a_i]}$ provides $U_\mathcal{P} = \sum_{i=1}^{n} M_i \Delta_i = \int_{[a,b]} u_\mathcal{P} dm$, and similarly for the lower sums. Hence for any *refining* sequence $(\mathcal{P}_n)$ of partitions (i.e. where $\mathcal{P}_n \subset \mathcal{P}_{n+1}$ for each n) with *mesh* $\rho^{(n)} = \max_{i \leq n} \Delta_i \to 0$, we obtain $l_{\mathcal{P}_n} \leq f \leq u_{\mathcal{P}_n}$. Thus on the measure space $([a, b], \mathcal{L}[a, b], m)$ – i.e., with Lebesgue measure restricted to the Lebesgue-measurable subsets of $[a, b]$– we have a monotone decreasing sequence $(u_{\mathcal{P}_n})_n$ with $u = \inf_n u_{\mathcal{P}_n} \geq f$ and a monotone increasing sequence $(l_{\mathcal{P}_n})_n$ with $l = \sup_n l_{\mathcal{P}_n} \leq f$. Both u and l are pointwise limits of their respective sequences, and all the functions in question are bounded by $M = \sup_{x \in [a,b]} f(x)$, which is in $\mathcal{L}^1([a, b])$. So by the DCT we have $\lim_n U_{\mathcal{P}_n} = \int_{[a,b]} u dm$ and $\lim_n L_{\mathcal{P}_n} = \int_{[a,b]} l dm$ and the limit functions u, l are in $\mathcal{L}^1([a, b])$.

Exercise 4.24 Prove that if x is not an endpoint of a partition interval (this excludes countably many points), then f is continuous at x iff $u(x) = f(x) = l(x)$.

Now if f is Riemann-integrable, we have $\int_{[a,b]} u dm = \int_{[a,b]} l dm$ and then $\int_a^b f(x) dx$ is their common value. But $l \leq f \leq u$ so $\int_{[a,b]} f dm$ also equals this common value, hence $u = f = l$ a.e.(m). Thus f is continuous a.e. (m) by the above Exercise. Conversely, if f is a.e.(m) continuous on $[a, b]$, then $u = f = l$ a.e.(m) and since u is measurable, so is f (this holds as m is a complete measure), and f is bounded, hence in $\mathcal{L}^1([a, b])$. For the Lebesgue integrals, we have

$$\int_{[a,b]} l dm = \int_{[a,b]} f dm = \int_{[a,b]} u dm.$$

The outer two are limits of Riemann sums, so f is also Riemann-integrable. To summarise: f is Riemann-integrable iff it is a.e.(m) continuous. In this event, its Riemann and Lebesgue integrals are equal.

Exercise 4.25 The *Dirichlet function* $f : [0, 1] \to [0, 1]$ has $f(x) = \frac{1}{n}$ whenever $x = \frac{m}{n}$ for some $m \leq n$ in $\mathbb{N}$, and is 0 otherwise. Show that

f is continuous at each irrational point, hence Riemann-integrable. (The last claim also follows from Riemann's integrability criterion, i.e. for all $\varepsilon > 0$ we can find a partition whose upper and lower sums are less than ε apart. Try proving this directly!)

Exercise 4.26

(i) Use the DCT to show that $\lim_{n\to\infty} \int_a^\infty \frac{n^2 x \exp(-n^2 x^2)}{1+x^2}\,dx = 0$ when $a > 0$. What is this limit when $a = 0$?

(ii) Use the Beppo-Levi Theorem to calculate $\int_0^1 \left(\frac{\log x}{1-x}\right)^2 dx$. (Hint: recall that $\sum_{n=1}^\infty \frac{1}{n^2} = \frac{\pi^2}{6}$.)

4.4 Product measures

We briefly consider the construction of product measures and integrals with respect to these. The first step is trivial: given two measurable spaces $(\Omega_i, \mathcal{F}_i)$, $i = 1, 2$, define the *product σ-field* $\mathcal{F} = \mathcal{F}_1 \times \mathcal{F}_2$ on the Cartesian product $\Omega = \Omega_1 \times \Omega_2$ as $\sigma(\mathcal{R})$ where $\mathcal{R} = \{F_1 \times F_2 : F_i \in \mathcal{F}_i, i = 1, 2\}$. (Think of sets in $\mathcal{R}$ as 'rectangles' by analogy with $\mathbb{R}^2 = \mathbb{R} \times \mathbb{R}$ and Lebesgue measure!) As $F_1 \times F_2 = (F_1 \times \Omega_2) \cap (\Omega_1 \times F_2)$, we can generate $\mathcal{F}$ from the class of 'cylinders' of the form $\{F_1 \times \Omega_2 : F_1 \in \mathcal{F}_1\}$ together with $\{\Omega_1 \times F_2 : F_2 \in \mathcal{F}_2\}$. Hence $\mathcal{F}$ is the smallest σ-field for which the two *projection maps* $\rho_i : \Omega_1 \times \Omega_2 \mapsto \Omega_i$, given by $\rho_i(\omega_1, \omega_2) = \omega_i$ $(i = 1, 2)$ are measurable. We have $\rho_1^{-1}(F_1) = F_1 \times \Omega_2$ and similarly for ρ_2.

Now $\mathcal{R}$ is clearly a π-system, and to extend results from rectangles to arbitrary sets in $\mathcal{F}$ we need a version of Theorem 2.8. Since we wish to integrate functions, we look to extend results that hold for indicators to general (bounded) measurable functions, so it will be useful to have this result in 'functional' form:

Theorem 4.27 (*Monotone Class Theorem*) *Suppose $\mathcal{H}$ is a vector space of bounded functions $\Omega \to \mathbb{R}$ with $1 \in \mathcal{H}$ and such that if $(f_n)_n$ are non-negative elements of $\mathcal{H}$ such that $f = \sup_n f_n$ is bounded, then $f \in \mathcal{H}$. If $\mathcal{H}$ contains all indicators of sets in a π-system $\mathcal{C}$, then $\mathcal{H}$ contains all bounded $\sigma(\mathcal{C})$-measurable functions.*

Proof The collection $\mathcal{D} = \{A \subset \Omega : \mathbf{1}_A \in \mathcal{H}\}$ is a d-system (because of the properties of $\mathcal{H}$) and contains $\mathcal{C}$, hence also contains $\sigma(\mathcal{C})$. Given a $\sigma(\mathcal{C})$-measurable function f with $0 \leq f(\omega) \leq K$ for all $\omega \in \Omega$ and some integer K, approximate f from below by simple functions $\phi_n(\omega) = \sum_{i=1}^{K2^n} \frac{i}{2^n} \mathbf{1}_{F_{n,i}}$ where $F_{n,i} = \left\{ \frac{i-1}{2^n} \leq f < \frac{i}{2^n} \right\}$ for each n. (Compare with Proposition 3.16.) Each $F_{n,i} \in \sigma(\mathcal{C})$, hence $\mathbf{1}_{F_{n,i}}$, and therefore each ϕ_n, is in $\mathcal{H}$. But then so is $f = \sup_n \phi_n$. For general bounded $\sigma(\mathcal{C})$-measurable f, apply this separately to f^+ and f^-.

Corollary 4.28 *Let $\mathcal{H}$ be class of bounded $\mathcal{F}$-measurable functions $f : \Omega_1 \times \Omega_2 \mapsto \mathbb{R}$ such that the map $\omega_1 \mapsto f(\omega_1, \omega_2)$ (resp. $\omega_2 \mapsto f(\omega_1, \omega_2)$) is $\mathcal{F}_2$- (resp. $\mathcal{F}_1$-)measurable for each $\omega_2 \in \Omega_2$ (resp. $\omega_1 \in \Omega_1$). Then $\mathcal{H}$ contains every bounded $\mathcal{F}$-measurable function.*

Proof Apply the above Monotone Class Theorem to the π-system $\mathcal{R}$ of rectangles. For $F = F_1 \times F_2 \in \mathcal{R}$, we have $\mathbf{1}_F \in \mathcal{H}$ and the conditions of the theorem are easily checked. Since $\mathcal{F} = \sigma(\mathcal{C})$, the Corollary follows.

Now suppose that $(\Omega_i, \mathcal{F}_i, P_i)$, $(i = 1, 2)$ are *probability* spaces. We wish to define the *product measure* $P = P_1 \times P_2$ on $(\mathbf{\Omega}, \mathcal{F})$ such that $P(A_1 \times A_2) = P(A_1)P(A_2)$ for $A_i \in \mathcal{F}_i, i = 1, 2$. With the Monotone Class Theorem to hand, it is convenient to construct P directly via integrals relative to P_1 and P_2: with $f \in \mathcal{H}$ as in the Corollary (i.e. the class of *bounded* $\mathcal{F}$-measurable functions on $\mathbf{\Omega}$) we can form the integrals $I_1^f(\omega_1) = \int_{\Omega_2} f(\omega_1, \omega_2 P_2(d\omega_2)$, and $I_2^f(\omega_2) = \int_{\Omega_1} f(\omega_1, \omega_2 P_1(d\omega_1)$ and another monotone class argument shows that I_i^f is $\mathcal{F}_i$-measurable $(i = 1, 2)$ and $\int_{\Omega_1} I_1^f dP_1 = \int_{\Omega_2} I_2^f dP_2$. This allows us to *define* $P = P_1 \times P_2$ for $F \in \mathcal{F}$ as the common value of these integrals when $f = \mathbf{1}_F$ (which clearly belongs to $\mathcal{H}$):

Theorem 4.29 *(Fubini) The map $P : \mathcal{F} \to [0, 1]$ defined for $F \in \mathcal{F}$ by*

$$P(F) = \int_{\Omega_1} I_1^{\mathbf{1}_F} dP_1 = \int_{\Omega_2} I_2^{\mathbf{1}_F} dP_2$$

is the unique probability measure on $(\mathbf{\Omega}, \mathcal{F})$ such that for A_i in $\mathcal{F}_i$ $(i = 1, 2)$, $P(A_1 \times A_2) = P(A_1)P(A_2)$. For every non-negative $\mathcal{F}$-measurable function $f : \mathbf{\Omega} \to \mathbb{R}$, we have

$$\int_{\boldsymbol{\Omega}} f \, dP = \int_{\Omega_1} \left[\int_{\Omega_2} f(\omega_1, \omega_2) P_2(d\omega_2) \right] P_1(d\omega_1)$$

$$= \int_{\Omega_2} \left[\int_{\Omega_1} f(\omega_1, \omega_2) P_1(d\omega_1) \right] P_2(d\omega_2).$$

The same holds if f is $\mathcal{F}$-measurable and $\int_{\boldsymbol{\Omega}} |f| \, dP < \infty$.

Proof Linearity and monotone convergence ensure that P is a measure. This is clearly a probability on $\boldsymbol{\Omega}$. For a rectangle $F = A_1 \times A_2$, we have

$$P(F) = \int_{\Omega_1} \left(\int_{\Omega_2} \mathbf{1}_{A_1 \times A_2} dP_2 \right) dP_1$$

$$= \int_{A_1} P(A_2) dP_1 = P(A_1) P(A_2).$$

Hence the iterated integrals are equal when $f = \mathbf{1}_{A_1 \times A_2}$. Again using the Monotone Class Theorem it follows that they are equal for all bounded $\mathcal{F}$-measurable f. This includes non-negative simple functions, and by the MCT also their increasing limits. Finally, if $|f|$ is P-integrable, we apply this separately to f^+, f^-. If another probability measure Q agrees with P on the π-system $\mathcal{R}$ of rectangles, then they agree on $\sigma(\mathcal{R}) = \mathcal{F}$. So P is the unique measure satisfying the requirement $P(A_1 \times A_2) = P(A_1)P(A_2)$.

The *product measure space* is now written as

$$(\boldsymbol{\Omega}, \mathcal{F}, P) = (\Omega_1, \mathcal{F}_1, P_1) \times (\Omega_2, \mathcal{F}_2, P_2).$$

The extension of this construction to n factors is clear. We may similarly define products of a sequence of probability spaces; however, when there are uncountably many factors things become markedly more complicated, as we shall see in Chapter 7.

4.5 Calculating expectations

When $(\Omega, \mathcal{F}, P)$ is a probability space, the integral $\int_{\Omega} X \, dP$ of a random variable $X : \Omega \to \mathbb{R}$ is its *expectation*, denoted by $\mathbb{E}[X]$. The map $X \mapsto \mathbb{E}(X)$ is linear, monotone and preserves constants ($\mathbb{E}[X] = c$ if $X(\omega) = c$ is constant). It is continuous under the restrictions given in the MCT and DCT. Note that $P(A) = \mathbb{E}[\mathbf{1}_A]$ for A in $\mathcal{F}$ and that

$\int_A X dP = \mathbb{E}[X \mathbf{1}_A]$. Moreover, Proposition 1.14, which showed that $A_n \to A$ implies $P(A_n) \to P(A)$, is now a special case of the DCT, applied to $\mathbf{1}_{A_N} \to \mathbf{1}_A$, since the $(\mathbf{1}_{A_n})_n$ are bounded by the (integrable!) constant function $\mathbf{1} = \mathbf{1}_\Omega$.

The distribution function F_X of X relates the abstract integral $\int_\Omega X dP$ to an integral on the line, and more generally:

Theorem 4.30 *If $X : \Omega \to \mathbb{R}$ is a random variable and g a real Borel function, then the random variable $Y = g(X)$ has expectation*

$$\mathbb{E}[Y] = \int_{\mathbb{R}} g(x) dF_X(x)$$

provided at least one of these quantities is finite. In particular, $Y \in \mathcal{L}^1(\Omega)$ iff $\int_{\mathbb{R}} |g(x)| dF_X(x) < \infty$.

Proof If $g = \mathbf{1}_B$ for B in $\mathcal{B}(\mathbb{R})$, both sides of the equation are $P(X^{-1}(B)) = P(X \in B)$. By linearity, they remain equal for any simple real function g, and since any *positive* Borel function g can be approximated from below by simple functions $g_n \uparrow g$, and thus $g_n(X) \uparrow g(X)$, the MCT (applied in $(\Omega, \mathcal{F}, P)$ and then in $(\mathbb{R}, \mathcal{B}, P_X)$, where $\int_{\mathbb{R}} g(x) dF_X(x) = \int_{\mathbb{R}} g dP_X$), ensures that

$$\mathbb{E}[g(X)] = \lim_n \mathbb{E}[g_n(X)] = \lim_n \int_{\mathbb{R}} g_n(x) dF_X(x)$$

$$= \int_{\mathbb{R}} g(x) dF_X(x).$$

For arbitrary Borel functions g, apply the above to g^+, g^- separately and subtract to conclude the proof.

In particular, we have $\mathbb{E}[g(X)] = \int_{\mathbb{R}} g(x) f_X(x) dm(x)$ if $F_X(t) = \int_{-\infty}^t f_X(u) dm(u)$ is an *absolutely continuous* distribution function (so that X has *density* f_X), and, at the other extreme

$$\mathbb{E}[g(X)] = \sum_i g(t_i) p_i$$

when X is *discrete*, with $p_i = P(X = t_i)$, $i \geq 1$.

Example 4.31 With $g(x) = x$, we obtain the *mean of* X, $\mathbb{E}[X] = \int_{\mathbb{R}} x dF_X(x)$, and similarly for the kth *moment:* $\mathbb{E}[X^k] =$

$\int_{\mathbb{R}} x^k dF_X(x)$. The *variance* of X is the second moment of the *centred* random variable $(X - \mathbb{E}[X])$, i.e.

$$Var(X) = \mathbb{E}[(X - \mathbb{E}[X])^2] = \int_{\mathbb{R}} (x - \mathbb{E}[X])^2 dF_X(x),$$

and the *standard deviation* of X, $\sigma_X = \sqrt{Var(X)}$, measures the dispersion of X about its mean.

Recall, from Example 3.22(ii), the *normal distribution*: we say that $X \sim \mathcal{N}(\mu, \sigma^2)$ if it has the density

$$f_X(x) = \frac{1}{\sqrt{2\pi\sigma^2}} \exp\left(-\frac{1}{2}\left[\frac{x - \mu}{\sigma}\right]^2\right).$$

In this case, the distribution F_X is determined by the parameters μ, σ, and it is routine to check that μ is the mean and σ^2 the variance of X. You may similarly compute the mean and variance of the other random variables described in Examples 1.5 and 3.22.

The next proposition compares the densities of X and Y when $Y = g(X)$ for a wide class of functions:

Proposition 4.32 *If $g : \mathbb{R} \to \mathbb{R}$ is increasing and differentiable (thus invertible), then*

$$f_{g(X)}(y) = f_X(g^{-1}(y))\frac{\mathrm{d}}{\mathrm{d}y}g^{-1}(y).$$

A similar result holds if g is decreasing.

Proof Consider the distribution function

$$F_{g(X)}(y) = P(g(X) \le y) = P(X \le g^{-1}(y)) = F_X(g^{-1}(y)).$$

Differentiate with respect to y to get the result. When g is decreasing the RHS is multiplied by (-1).

Example 4.33 If X is standard normal (i.e. with mean 0 and variance 1), then $Y = \mu + \sigma X$ is $Y \sim \mathcal{N}(\mu, \sigma^2)$. This follows at once: with $g^{-1}(y) = \frac{y-\mu}{\sigma}$, the derivative is $\frac{1}{\sigma}$. Similarly, when $g(x) = e^x$ we obtain the *log-normal* distribution, which is widely used in financial applications.

Exercise 4.34 The random variable X measures the distance from points in the unit square $[0, 1]^2$ to the nearest edge. Find its distribution and expectation.

The following two inequalities involving $\mathbb{E}[g(X)]$ for different classes of real functions g have applications in many varied settings. For the first, recall that a real function g, defined on a open interval I, is *convex* on I if its graph lies below any of its chords, i.e. for all x, y in I
$g(cx + (1 - c)y) \leq cg(x) + (1 - c)g(y)$ whenever $0 \leq c \leq 1$.

Proposition 4.35 *(Jensen's inequality) If g is convex and X and $g(X)$ are in $\mathcal{L}^1(\Omega)$, then $g(\mathbb{E}[X]) \leq \mathbb{E}[g(X)]$.*

Proof To prove this, note that for convex g the tangent to a point of its graph lies below the graph. Thus for $x < y$ the slope of the tangent at y exceeds that of the corresponding chord, i.e. with $a = g'(y)$ (or a one-sided derivative if necessary) we obtain $g(x) \geq g(y) + a(x - y)$. Using this with $y = \mathbb{E}[X]$ and $x = X(\omega)$ for $\omega \in \Omega$ and taking expectations yields Jensen's inequality.

As $g(x) = |x|$ is convex, we again have $|\mathbb{E}[X]| \leq \mathbb{E}[|X|]$.

Proposition 4.36 *(Markov's inequality) If $X \geq 0$ and g is positive and increasing on $[0, \infty)$, then for any $a > 0$*

$$P(X \geq a) \leq \frac{\mathbb{E}[g(X)]}{g(a)}.$$

Proof Now $g(X) \geq g(X)\mathbf{1}_{\{X \geq a\}} \geq g(a)\mathbf{1}_{\{X \geq a\}}$ and we take expectations.

The inequality provides a quantitative refinement of the obvious fact that an integrable random variable can only become large with small probability. In particular, with $g(x) = x^p$ for $p \geq 1$ and $|X|$ yields

$$P(|X| > a) \leq \frac{\mathbb{E}[|X|^p]}{a^p}.$$

With $p = 2$, this is known as *Chebychev's inequality*. In particular, if X has mean μ and variance σ^2, we use $X - \mu$, $p = 2$ and $a = k\sigma$ to obtain $P(|X - \mu| \geq k\sigma) \leq \frac{1}{k^2}$. Thus a random variable with small variance remains close to its mean with high probability.

Example 4.37 We close with two further useful formulae for the calculation of expectations:

(i) If $X \in \mathcal{L}^1_+(\Omega)$ – so that $X \geq 0$ a.e.(P) – then with $m = m_{[0,\infty)}$ as the Lebesgue measure restricted to $[0, \infty)$

$$\mathbb{E}[X] = \int_0^\infty P(X \geq x)dm(x) = \int_0^\infty [1 - F_X(x)]dm(x).$$

(ii) $\mathbb{E}[X] = \sum_{=1}^\infty nP(X = n) = \sum_{n=1}^\infty P(X \geq n)$ for $X : \Omega \to \mathbb{N}$.

To prove (i), use Fubini's Theorem on the product measure space $(\Omega \times [0, \infty), \mathcal{F} \times \mathcal{B}[0, \infty)\ P \times m)$ and consider $F = \{(\omega, x) : 0 \leq x \leq X(\omega)\}$ – this is just the region in the product space 'lying under the graph' of X. We have

$$I_1^{1_F}(\omega) = \int_0^\infty \mathbf{1}_F(\omega, x)dm(x) = X(\omega),$$

while

$$I_2^{1_F}(x) = \int_\Omega \mathbf{1}_F(\omega, x)P(d\omega) = P(X \geq x).$$

By Fubini

$$\mathbb{E}[X] = \int_\Omega I_1^{1_F}dP = \int_0^\infty I_2^{1_F}(x)dm(x)$$

$$= \int_0^\infty P(X \geq x)dm(x).$$

For the final identity, note that F_X is monotone, and hence has only countably many discontinuities, so that $P(X = x) > 0$ at only countably many $x \geq 0$, i.e. $P(X \geq x) = P(X > x)$ a.s. (m). For (ii), simply consider F_X on each $[n - 1, n]$.

5

L^p-spaces and conditional expectation

Let $(\Omega, \mathcal{F}, P)$ be a probability space and X a random variable (we restrict attention to this situation from now on). We adopt a slightly different point of view and examine X as a point in a *normed vector space* of functions. To do this, we shall need to identify (i.e. treat as the same) functions which are equal a.s.(P), since the integral cannot distinguish between them. We then examine the *inner product* structure of the space of square-integrable random variables, which will enable us to generalise the idea of the *conditional expectation* of X: for $B \in \mathcal{F}$ with $P(B) > 0$ this is defined by $\mathbb{E}[X|B] = \frac{1}{P(B)} \int_\Omega X \, dP$. However, in trying to define $\mathbb{E}[X|Y]$, where Y is another random variable, we hit a problem: the events $(Y = c)$ need not have positive probability for any real c (e.g. when Y has a density). We develop a different, more general, approach to resolve the matter.

5.1 L^p as a Banach space

For any $p \geq 1$, define $\mathcal{L}^p(\Omega) = \{X : \Omega \to \mathbb{R} : \mathbb{E}[|X|^p] < \infty\}$. We use the map $||\cdot||_p : X \to (\mathbb{E}[|X|^p])^{1/p}$ to define a *norm*. To achieve this, we must identify random variables that are equal a.s. (P), as $||X - Y||_p = 0$ iff $X = Y$ a.s.(P). Now define $X \sim Y$ iff $X = Y$ a.s.(P): this is an equivalence relation on $\mathcal{L}^p(\Omega)$ and we denote the quotient space $\mathcal{L}^p(\Omega)/ \sim$ by $L^p(\Omega)$. We call representatives of the same class *versions* of each other, and continue to write X, Y for members of $L^p(\Omega)$, i.e. we identify an equivalence class with any of its representatives.

To see that $(L^p(\Omega), ||\cdot||)$ is a normed vector space, note first that if X, Y are random variables, then so are $X + Y$, cX for real c. Since

49

$|cX|^p = |c|^p |X|^p$ and $|X + Y|^p \leq 2^p \max\{|X|^p, |Y|^p\}$, it follows that $L^p(\Omega)$ is a vector space and $||cX||_p = |c| \, ||X||_p$. Now $||X||_p = 0$ iff $X = 0$ (as an element of $L^p(\Omega)$!). So all that remains is to prove the triangle inequality for this norm.

First, we need an important consequence of Jensen's inequality:

Proposition 5.1 *(Hölder inequality) Let* $p > 1$, $\frac{1}{p} + \frac{1}{q} = 1$. *If* $X \in L^p(\Omega)$ *and* $Y \in L^q(\Omega)$, *then* $XY \in L^1(\Omega)$ *and*

$$|\mathbb{E}[XY]| \leq ||XY||_1 \leq ||X||_p \, ||Y||_q.$$

Proof We may assume without loss that X, Y are non-negative and $||X||_p > 0$. From Jensen's inequality applied to the convex function $x \to x^q$, we obtain for any $W \in L^q(\Omega, \mathcal{F}, Q)$

$$\left(\int_\Omega W \, dQ \right)^q \leq \int_\Omega W^q dQ.$$

Define a new probability Q on $\mathcal{F}$ by

$$Q(A) = \frac{1}{||X||_p^p} \int_A X^p dP \quad \text{(informally: } X^p dP = ||X||_p^p \, dQ\text{)}.$$

Now let $Z(\omega) = \frac{Y(\omega)}{X(\omega)^{p-1}}$ on $A = \{\omega : X(\omega) > 0\}$ and 0 on $\Omega \backslash A$, so that on A, $Y = ZX^{p-1}$ and $Z^q = \frac{Y^q}{X^p}$ since $p + q = pq$. Thus $\int_\Omega Z^q dQ = \frac{1}{||X||_p^p} \int_A Y^q dQ$ is finite. Now use the above inequality with $W = ||X||_p^p Z \mathbf{1}_A \in L^q(\Omega, \mathcal{F}, Q)$

$$\int_\Omega (XY) dP = \int_A X^p Z dP = \int_\Omega W \, dQ \leq \left[\int_\Omega W^q dQ \right]^{1/q}$$

$$= \left[\int_\Omega ||X_p||_p^{pq} \left(\frac{Y^q}{X^p} \right) \mathbf{1}_A \frac{X^p}{||X||_p^p} dP \right]^{1/q}$$

$$= ||X||_p \left[\int_\Omega Y^q \mathbf{1}_A dP \right]^{1/q} \leq ||X||_p \, ||Y||_q.$$

A familiar special case occurs when $p = q = 2$.

Corollary 5.2 *(Schwarz inequality) When* $X, Y \in L^2(\Omega)$, *then* $XY \in L^1(\Omega)$ *and* $|\mathbb{E}[XY]| \leq ||XY||_1 \leq ||X||_2 \, ||Y||_2$.

We can now prove the triangle inequality for the norm $X \to ||X||_p$ on L^p whenever $p \geq 1$.

Corollary 5.3 *(Minkowski's inequality) If $p \geq 1$ and $X, Y \in L^p(\Omega)$, then $||X + Y||_p \leq ||X||_p + ||Y||_p$.*

Proof The case $p = 1$ is trivial, so assume $p > 1$. Now

$$|X + Y|^p \leq |X| |X + Y|^{p-1} + |Y| |X + Y|^{p-1}$$

and with $\frac{1}{p} + \frac{1}{q} = 1$ we have $|X + Y|^{(p-1)q} = |X + Y|^p$, so that $|X + Y|^{p-1} \in L^q$. Apply Hölder's inequality to each product

$$\int_\Omega |X| |X + Y|^{p-1} \, dP \leq ||X||_p \, C,$$

where $C = [\int_\Omega (|X + Y|^{p-1})^q dP]^{1/q} = ||X + Y||_p^{p/q}$, and the same remains true with X, Y interchanged. Hence

$$||X + Y||_p^p \leq C(||X||_p + ||Y||_p).$$

Dividing both sides by C completes the proof, as $p - \frac{p}{q} = 1$.

Remark 5.4 Note that the above proofs do not use the finiteness of P: the Jensen inequality was used for Q, which is a probability by its definition. Thus the above inequalities are valid for any measure space. For our next result, however, the finiteness of P is essential.

Proposition 5.5 *The L^p-norm is monotone in p when $P(\Omega)$ is finite: when $1 \leq p \leq q < \infty$, then $L^q \subset L^p$ and for any $X \in L^q$ we have $||X||_p \leq ||X||_q$.*

Proof We use the Hölder inequality, which holds for $r > 1$ and $\frac{1}{r} + \frac{1}{s} = 1$ to yield $\mathbb{E}[|YZ|] \leq (\mathbb{E}[|Y|^r])^{1/r}(\mathbb{E}[|Z|^s])^{1/s}$. In particular, for $Z = 1$ (constant, hence integrable, as $P(\Omega)$ is finite!) this reads: $\mathbb{E}[|Y|] \leq (\mathbb{E}[|Y|^r])^{1/r}$. Now, for $1 \leq p \leq q < \infty$ set $r = \frac{q}{p}$ and let $Y = |X|^p$. We obtain $\mathbb{E}[|X|^p] \leq (\mathbb{E}[|X|^{pr}])^{1/r}) = (\mathbb{E}[|X|^q])^{p/q}$ and taking pth roots on both sides gives the result.

Exercise 5.6 Use the Jensen inequality, with $g(x) = x^{q/p}$, to prove this inequality directly.

Example 5.7 Proposition 5.5 shows in particular that if the random variable has finite kth moment ($\mathbb{E}[|X|^k] < \infty$), then for all $n < k$ the nth moment is also finite; while if the kth moment is infinite, so are all higher moments.

Remark 5.8 The vector space $L^\infty(\Omega)$ of essentially bounded random variables (see Definition 3.14) – where we again identify random variables that are equal a.s.(P) – has the norm $||X||_\infty = \mathrm{ess\,sup}_{\omega \in \Omega} |X(\omega)|$ (check that it is a norm!). The (much simpler) proof for the case $p = \infty$ is left for you.

Remark 5.9 The vector space of all (a.s. finite) random variables $X : \Omega \to \mathbb{R}$ is sometimes denoted by $\mathcal{L}^0(\mathcal{F})$ and if we again identify X and Y when $X = Y$ a.s.(P), we obtain the complete metric space $L^0(\mathcal{F})$ via the metric $d(X,Y) = \int_\Omega \min(1, |X - Y|)dP$. The metric is not, however, induced by a norm on this space.

Recall that a normed space $(E, ||\cdot||)$ is *complete* (alternatively, is a *Banach space*) if every Cauchy sequence in E converges to an element of E, i.e. $||x_m - x_n|| \to 0$ when $m, n \to \infty$ implies that $||x_n - x|| \to 0$ for some $x \in E$ as $n \to \infty$. The L^p-spaces provide important examples of Banach spaces; again we state the result for any E in $\mathcal{F}$:

Theorem 5.10 *The normed vector space $L^p(E)$ is complete for $1 \le p \le \infty$.*

Proof (Sketch) For a Cauchy sequence X_n, find a subsequence X_{n_k} with $||X_{k_{n+1}} - X_{k_n}||_p < \frac{1}{2^k}$ for all $n \ge 1$ and note that by Proposition 5.5

$$\mathbb{E}[|X_{k_{n+1}} - X_{k_n}|] \le ||X_{k_{n+1}} - X_{k_n}||_p < \frac{1}{2^n}$$

for each n. Hence $\mathbb{E}\left[\sum_{n=1}^\infty |X_{k_{n+1}} - X_{k_n}|\right]$ is finite, so the series $\sum_n (X_{k_{n+1}} - X_{k_n})$ converges absolutely, so that the sequence $(X_{k_n})_n$ converges a.s.(P). Let $X = \limsup_n X_{k_n}$, which is measurable, and take $r > k_n$. For all $m \ge n$, $\mathbb{E}[|X_r - X_{k_m}|^p] \le \frac{1}{2^{pn}}$, and as $m \uparrow \infty$ Fatou's Lemma yields $\mathbb{E}[|X_r - X|^p] \le \frac{1}{2^{pn}}$. Thus $(X_r - X)$ and X are in L^p and $||X_r - X||_p \to 0$ as $r \to \infty$.

Exercise 5.11 Show that the space $L^\infty(E)$ is complete with the norm $||X||_\infty = ess \sup_{\omega \in \Omega} |X(\omega)|$.

For later reference, we derive a useful consequence of Hölder's inequality, coupled with Fubini's Theorem:

Proposition 5.12 *Suppose $p > 1$, $\frac{1}{p} + \frac{1}{q} = 1$, and X, Y are nonnegative random variables with $Y \in L^p(\Omega)$, and*

$$\lambda P(X \geq \lambda) \leq \int_{\{X \geq \lambda\}} Y dP$$

for all $\lambda \geq 0$. Then $X \in L^p$ and $||X||_p \leq q||Y||_p$.

Proof Fix $n > 0$ and let $X_n = X \wedge n$. Then X_n is bounded, hence in L^p. If we have proved our inequality for Y and X_n, the MCT shows that it also holds for Y and $X = \lim_{n \uparrow \infty} X_n$. We can thus take $X \in L^p$ without loss.

Since for any $z \geq 0$, $p \int_{\{z \geq x\}} x^{p-1} dx = p \int_0^z x^{p-1} dx = z^p$, we have, integrating $z = X(\omega)$, over Ω and using Fubini

$$\int_\Omega X^p dP = \int_\Omega p \left[\int_0^{X(\omega)} x^{p-1} dx \right] P(d\omega)$$

$$= p \int_0^\infty \left[\int_\Omega 1_{\{X(\omega) \geq x\}} P(d\omega) \right] x^{p-1} dx$$

$$= p \int_0^\infty x^{p-1} P(X \geq x) dx.$$

But $x P(X \geq x) \leq \int_{\{X \geq x\}} Y dP$ by hypothesis, so (using Fubini again) the RHS is no greater than

$$p \int_0^\infty x^{p-2} \int_{\{X \geq x\}} Y dP dx = p \int_\Omega Y \int_0^\infty x^{p-2} 1_{\{X \geq x\}} dx dP.$$

With $\frac{1}{p} + \frac{1}{q} = 1$, $p = q(p - 1)$ the RHS equals

$$q \int_\Omega Y(\omega) \left[\int_0^{X(\omega)} (p-1) x^{p-2} dx \right] P(d\omega) = q\mathbb{E}[Y X^{p-1}].$$

Since $X^{p-1} \in L^q$, Hölder's inequality now provides that $\mathbb{E}[Y X^{p-1}] \leq ||X^{p-1}||_q ||Y||_p$. The result follows, as $(p - 1)q = p$, so that $||X^{p-1}||_q = (\int_\Omega X^p dP)^{1/q}$.

5.2 Orthogonal projections in L^2

When $p = 2$, the norm $X \to ||X||_2 = (\mathbb{E}[X^2])^{1/2}$ (we omit the modulus as X is real-valued) is given by the *inner product* $(X, Y) = \mathbb{E}[XY]$. First, note that the Schwarz inequality yields $XY \in L^1$ if $X, Y \in L^2$, so that the inner product is well-defined. We check that it is an inner product: it is *linear* since $(aX + bY, Z) = a(X, Z) + b(Y, Z)$, *symmetric* as $(Y, X) = (X, Y)$ and *positive-definite* since $(X, X) \geq 0$ and $= 0$ iff $X = Y$ a.s.(P). Thus by Theorem 5.10, $(L^2(\Omega), (\cdot, \cdot))$ is complete inner product space, i.e. a *Hilbert space*.

The *covariance* of random variables X, Y in L^2 with means μ_X, μ_Y is the inner product of their *centred* versions: $Cov(X, Y) = \mathbb{E}[X_c Y_c] = \mathbb{E}[XY] - \mu_X \mu_Y$. (We write $Z_c = Z - \mu_Z$ for any Z.) The *variance* σ_X^2 of X is the special case when $Y = X$.

If $Cov(X, Y) = 0$, we say that X, Y are *uncorrelated*. This just expresses the *orthogonality* of X_c and Y_c in L^2. For X, Y we then have $\mathbb{E}[XY] = \mathbb{E}[X]\mathbb{E}[Y]$. More generally, the *correlation coefficient* $\rho_{X,Y}$ between non-zero X and Y in L^2 is given by $\rho_{X,Y} = \cos \theta = \frac{Cov(X,Y)}{\sigma_X \sigma_Y}$. $\rho_{X,Y}$ is therefore the cosine of the *angle* between X and Y. By Schwarz, $|\rho_{X,Y}| \leq 1$. Also $|\rho_{X,Y}| = 1$ iff equality holds in Schwarz iff the vectors X_c, Y_c are *linearly* dependent (for some real a, b, $P(aX_c + bY_c = 0) = 1$.)

Proposition 5.13 *For random variables* $X_1, X_2, \ldots, X_n$ *in* $L^2(\Omega)$ *and real numbers* $a_1, a_2, \ldots, a_n$ *we have*

$$Var \left(\sum_{i=1}^{n} a_i X_i \right) = \sum_{i=1}^{n} a_i^2 Var(X_i) + 2 \sum_{i \neq j} a_i a_j Cov(X_i, X_j).$$

If the (X_i) *are uncorrelated,* $Var \left(\sum_{i=1}^{n} X_i \right) = \sum_{i=1}^{n} Var(X_i)$.

This follows at once from the definitions, and is left as an easy exercise.

Independent random variables are always uncorrelated, as the next Exercise shows.

Exercise 5.14   Prove carefully that X, Y are independent iff $\mathbb{E}[f(X)g(Y)] = \mathbb{E}[f(X)]\mathbb{E}[g(Y)]$ for all bounded Borel functions f, g. (*Hint*: begin with indicators and use linearity and dominated convergence.) Use this result to show that if X, Y are independent, then

$\mathbb{E}[XY] = \mathbb{E}[X]\mathbb{E}[Y]$. (*Caution!* The function $f(x) = x$ is not bounded, so you need truncation and limit arguments here!)

Remark 5.15 However, uncorrelated random variables need not be independent. You may check that if $X = \cos\theta$ and $Y = \sin\theta$ with θ uniformly distributed on $[0, 2\pi]$, then X, Y are uncorrelated but not independent. On the other hand, *normally* distributed random variables are independent iff they are uncorrelated – this follows at once from the form of the normal density.

Geometrically, the bilinearity of the inner product in L^2 has the following familiar consequences – their easy proofs are left for you!
(i) **Pythagoras' Theorem:** If $(X, Y) = 0$, then

$$||X + Y||_2^2 = ||X||_2^2 + ||Y||_2^2.$$

(ii) **Parallelogram Law:** For any X, Y in L^2

$$||X + Y||_2^2 + ||X - Y||_2^2 = 2\left(||X||_2^2 + ||Y||_2^2\right).$$

Less obvious is that the parallelogram law *characterises* inner product norms: if it fails, there is no inner product that can induce the given norm. We shall not prove this, but note the following exercise:

Exercise 5.16 Show that the parallelogram law fails for the normed space $L^1([0, 1], \mathcal{B}[0, 1], m)$. (*Hint:* Consider the functions $X(x) = \frac{1}{2} - x, Y(x) = x - \frac{1}{2}$.)

A vector subspace $\mathcal{H}$ of $L^2(\Omega)$ is *complete* if any norm-Cauchy sequence in $\mathcal{H}$ converges to an element of $\mathcal{H}$ (we know that it must converge to some vector in L^2, since L^2 is complete – the additional requirement is that the limit is also in $\mathcal{H}$). For such subspaces, there are well-defined orthogonal projections:

Theorem 5.17 *Let $X \in L^2(\Omega)$ and suppose $\mathcal{K}$ is a complete subspace of $L^2(\Omega)$. Then we can find a unique (a.s.(P)) Y in $\mathcal{K}$ such that $X - Y$ is orthogonal to $\mathcal{K}$, i.e. $(X - Y, Z) = 0$ for all Z in $\mathcal{K}$).*

Proof We shall exhibit Y as the point of $\mathcal{K}$ 'nearest' to X in that
$\delta_{\mathcal{K}} = ||X - Y||_2 = \inf\{||X - Z||_2 : Z \in \mathcal{K}\}.$

First, choose a sequence (Y_n) in $\mathcal{K}$ such that $||X - Y_n||_2 \to \delta_K$. The parallelogram law, applied to $Z = X - \frac{1}{2}(Y_n + Y_m)$ and $W = \frac{1}{2}(Y_n - Y_m)$, noting that $\frac{1}{2}(Y_n \pm Y_m) \in \mathcal{K}$, gives

$$||X - Y_m||_2^2 + ||X - Y_n||_2^2 = 2\left(||Z||_2^2 + ||W||_2^2\right).$$

Now, as $m, n \to \infty$, both terms on the left and the first on right converge to δ_K^2, and hence $||W||_2^2 \to 0$. This shows that $(Y_n)_n$ is Cauchy, and so converges to a unique element Y of $\mathcal{K}$. Thus we have $||X - Y||_2 = \delta_K$ and the same holds for any random variable Y' with $Y' = Y$ a.s.(P). Now note that for any real t and $Z \in \mathcal{K}$ we have $Y + tZ \in \mathcal{K}$, so that $||X - (Y + tZ)||_2^2 \ge ||X - Y||_2^2$, and multiplying out the inner products yields $t^2||Z||_2^2 \ge 2t(X - Y, Z)$. By taking $t > 0$ small enough, we see that this can only remain true for all t if $(X - Y, Z) = 0$.

Remark 5.18 Pythagoras' Theorem shows that, conversely, if we have Y in $\mathcal{K}$ with $(X - Y)$ orthogonal to $\mathcal{K}$, then $||X - Y||_2 = \delta_K$. The details are left to you.

5.3 Properties of conditional expectation

The simple definition of conditional expectation applies to conditioning on discrete random variables: if $Y \in L^1(\Omega)$ takes values $(y_i)_{i \ge 1}$ with $B_i = (Y = y_i)$ and $P(B_i) > 0$ for each i, then these (disjoint) events *partition* Ω and for any integrable X we can define the *random variable* $\mathbb{E}[X|Y]$ by $\mathbb{E}[X|Y](\omega) = \mathbb{E}[X|B_i]$ for $\omega \in B_i$. Now the partition $(B_i)_i$ generates the σ-field $\sigma(Y)$ (see Section 3.3) – in fact, any $A \in \sigma(Y)$ is a countable union of sets B_i. But $\mathbb{E}[X|Y]$ is constant on each B_i, so it is $\sigma(Y)$-measurable (i.e. $\mathbb{E}[X|Y]^{-1}(B) \in \sigma(Y)$ for each Borel set B in $\mathbb{R}$). For $i \ge 1$, $\int_{B_i} \mathbb{E}[X|Y]dP = \int_{B_i} \mathbb{E}[X|B_i]dP = \int_{B_i} XdP$. Summing over the i such that $B_i \subset A$ we have, for all A in $\sigma(Y)$, $\int_A \mathbb{E}[X|Y]dP = \int_A XdP$.

These two properties *characterise* the conditional expectation $\mathbb{E}[X|Y]$ when Y is discrete. For general Y, we use them as the definition:

Definition 5.19 If $X \in L^1(\Omega)$ and Y is any random variable (a version of), the *conditional expectation of X given Y* is a $\sigma(Y)$-measurable random variable $\mathbb{E}[X|Y]$ such that $\int_A \mathbb{E}[X|Y]dP = \int_A XdP$ for all A in $\sigma(Y)$.

We have not yet shown that such a thing always exists; but if it does, it must be unique a.s.(P); in fact:

Lemma 5.20 *If $\mathcal{G}$ is a sub-σ-field of $\mathcal{F}$ and Z is $\mathcal{G}$-measurable with $\int_G ZdP = 0$ for all G in $\mathcal{G}$, then $Z = 0$ a.s.(P).*

Proof For any $n \geq 1$, the event $G_n = \left(Z \geq \frac{1}{n}\right) \in \mathcal{G}$ is P-null, since $0 \leq \frac{1}{n}P(G_n) \leq \int_{G_n} ZdP = 0$. Similarly, for $\left(Z \leq -\frac{1}{n}\right)$, so that $P(B_n) = 1$ when $B_n = \left(|Z| \leq \frac{1}{n}\right)$. But $B_n \downarrow (Z = 0)$, so $P(Z=0) = \lim_n P(B_n) = 1$.

Hence if $X_1 = X_2$ a.s.(P), so that for all G in $\sigma(Y)$

$$0 = \int_G (X_1 - X_2)dP = \int_G (\mathbb{E}[X_1|Y] - \mathbb{E}[X_2|Y])dP,$$

then $\mathbb{E}[X_1|Y] = \mathbb{E}[X_2|Y]$ a.s. (P).

The lemma shows that what really matters here is $\sigma(Y)$ rather than the values of Y. In fact, what we showed was that if random variables Y_1, Y_2 generate the same σ-field ($\sigma(Y_1) = \sigma(Y_2)$), then $\mathbb{E}[X|Y_1] = \mathbb{E}[X|Y_2]$ for any integrable X. This motivates our final definition:

Definition 5.21 If $X \in L^1(\Omega, \mathcal{F}, P)$ and $\mathcal{G}$ is a sub-σ-field of $\mathcal{F}$, the *conditional expectation of X given $\mathcal{G}$* is the $\mathcal{G}$-measurable random variable $\mathbb{E}[X|\mathcal{G}]$ such that $\int_G \mathbb{E}[X|\mathcal{G}]dP = \int_G XdP$ for all $G \in \mathcal{G}$.

For the conditional probability, we have $P(F|\mathcal{G}) = \mathbb{E}[\mathbf{1}_F|\mathcal{G}]$.

The existence of $\mathbb{E}[X|\mathcal{G}]$ for any $X \in L^2(\mathcal{F})$ is now clear – it is the orthogonal projection $X_\mathcal{G}$ of X onto the complete subspace $L^2(\mathcal{G})$ – to see that it satisfies our definition, note that $X_\mathcal{G}$ is $\mathcal{G}$-measurable and that $(X - X_\mathcal{G}, Z) = 0$ for all $Z \in L^2(\mathcal{G})$. But this says that $\int_G XdP = \int_G X_\mathcal{G}dP$ for all G in $\mathcal{G}$. So we write $\mathbb{E}[X|\mathcal{G}]$ for $X_\mathcal{G}$ from now on.

Remark 5.22 It is possible to extend the construction of $\mathbb{E}[X|\mathcal{G}]$ to all X in $L^1(\Omega)$ by a careful limit argument (see [W], [C-K]), but we shall take this for granted now and return to it in the next chapter (Theorem 6.36) in a more general setting. Our proof of the uniqueness holds in L^1 also.

Assuming existence, we derive the main properties of conditional expectation for integrable random variables.

Proposition 5.23 *Suppose* $X, Y \in L^1(\Omega, \mathcal{F}, P)$, $a, b \in \mathbb{R}$ *and* $\mathcal{H} \subset \mathcal{G}$ *are sub-σ-fields of* $\mathcal{F}$. *Then we have:*

(i) (Linearity) $\mathbb{E}[(aX + bY)|\mathcal{G}] = a\mathbb{E}[X|\mathcal{G}] + b\mathbb{E}[Y|\mathcal{G}]$ *(more precisely: a linear combination of versions of* $\mathbb{E}[X|\mathcal{G}]$ *and* $\mathbb{E}[Y|\mathcal{G}]$ *is a version of the conditional expectation of the same linear combination of* X *and* Y).

(ii) $\mathbb{E}[\mathbb{E}[X|\mathcal{G}]] = \mathbb{E}[X]$ *(thus any version of* $\mathbb{E}[X|\mathcal{G}]$ *has the same expectation as* X).

(iii) If X *is* $\mathcal{G}$-*measurable*, $\mathbb{E}[X|\mathcal{G}] = X$ *a.s.*(P) *(more precisely: X is a version of* $\mathbb{E}[X|\mathcal{G}]$).

(iv) (Independence) If $\sigma(X)$ *and* $\mathcal{G}$ *are independent, then* $\mathbb{E}[X|\mathcal{G}] = \mathbb{E}[X]$ *(i.e. it is a.s. constant).*

(v) (Positivity) If $X \geq 0$ *a.s.*(P), *the same holds for any version of* $\mathbb{E}[X|\mathcal{G}]$.

(vi) (Monotone convergence) If $X_n \uparrow X$ *in* L^1_+, *then* $\mathbb{E}[X_n|\mathcal{G}] \uparrow \mathbb{E}[X|\mathcal{G}]$ *(both hold a.s.*(P)).

(vii) (Tower property) $\mathbb{E}[(\mathbb{E}[X|\mathcal{G}])|\mathcal{H}] = \mathbb{E}[X|\mathcal{H}]$ *a.s.*(P).

(viii) ('Taking out what is known') If Y *is* $\mathcal{G}$-*measurable and* $XY \in L^1(\Omega, \mathcal{F}, P)$, *then* $\mathbb{E}[(XY)|\mathcal{G}] = Y\mathbb{E}[X|\mathcal{G}]$ *a.s.* (P).

(ix) (Jensen's inequality) If g *is a convex real function and* $g(X) \in L^1$, *then* $g(\mathbb{E}[X|\mathcal{G}]) \leq \mathbb{E}[g(X)|\mathcal{G}]$.

(x) The map $X \to \mathbb{E}[X|\mathcal{G}]$ *contracts the* L^p-*norm for each* $p \geq 1$: $||\mathbb{E}[X|\mathcal{G}]||_p \leq ||X||_p$.

Proof (i)–(iv) follow immediately from the definitions and are left as exercises. For (v), $G_k = (\mathbb{E}[X|\mathcal{G}] < -\frac{1}{k})$ is in $\mathcal{G}$, so from $0 \leq \int_{G_k} X dP = \int_{G_k} \mathbb{E}[X|\mathcal{G}] dP \leq -\frac{1}{k} P(G_k)$ we conclude that $P(\mathbb{E}[X|\mathcal{G}] < 0) = P(\cup_k G_k) = 0$. So any version of $\mathbb{E}[X|\mathcal{G}]$ is a.s.(P) non-negative.

For (vi), let Y_n be a version of $\mathbb{E}[X_n|\mathcal{G}]$. These are non-negative and increase to $\mathbb{E}[X|\mathcal{G}]$ a.s.(P), since $(\int_G X_n dP)_n$ is increasing for each G in $\mathcal{G}$. Take $Y = \limsup_n Y_n$, which is the $\mathcal{G}$-measurable a.s(P)-limit of $(Y_n)_n$, so that for G in $\mathcal{G}$, $\int_G X_n dP = \int_G Y_n dP \uparrow \int_G Y dP$. But the LHS (left-hand side) increases to $\int_G X dP$ by the MCT, so Y is a version of $\mathbb{E}[X|\mathcal{G}]$.

(vii) is a consequence of the definition: $\int_G \mathbb{E}[X|\mathcal{G}]dP = \int_G XdP$ holds on $\mathcal{G}$ hence on $\mathcal{H}$. Now for $H \in \mathcal{H}$ we have $\int_H \mathbb{E}[X|\mathcal{H}]dP = \int_H XdP$, so also $\int_H \mathbb{E}[X|\mathcal{G}]dP = \int_H \mathbb{E}[X|\mathcal{H}]dP$. In other words, the $\mathcal{H}$-measurable $\mathbb{E}[X|\mathcal{H}]$ is a version of $\mathbb{E}[(\mathbb{E}[X|\mathcal{G}])|\mathcal{H}]$.

For (viii), we restrict to $X \geq 0$ and then use (i) applied to $X^+ - X^-$. When $Y = \mathbf{1}_E$ and $G, E \in \mathcal{G}$, then

$$\int_G \mathbf{1}_E \mathbb{E}[X|\mathcal{G}]dP = \int_{E \cap G} XdP = \int_G \mathbf{1}_E XdP.$$

So $\mathbf{1}_E \mathbb{E}[X|\mathcal{G}]$ is a version of $\mathbb{E}[(XY)|\mathcal{G}]$. Linearity and (vi) applied to a sequence of $\mathcal{G}$-simple functions increasing to Y does the rest.

To prove (ix), we use without proof (see [W]) that a convex function g is the supremum of a sequence of affine functions, i.e. $g(x) = \sup(a_n x + b_n)$ for some real sequences $(a_n), (b_n)$. Now for each fixed n, $g(X(\omega))$ dominates $a_n X(\omega) + b_n$ a.s.(P), and by (v) the same holds with X replaced by any version Y of $\mathbb{E}[X|\mathcal{G}]$. The union of the sets A_n where this fails is P-null, so $\mathbb{E}[g(X)|\mathcal{G}] \geq \sup_n a_n \mathbb{E}[X|\mathcal{G}] + b_n = g(\mathbb{E}[X|\mathcal{G}])$ holds a.s. (P).

(x) follows from (ix) and (i), as the map $x \to |x|^p$ is convex for $p \geq 1$.

Exercise 5.24 Let $\Omega = [0,1]$ with Lebesgue measure and suppose $X(\omega) = \omega$. If $0 < a < 1$, compute $\mathbb{E}[X|\mathcal{G}]$ when $\mathcal{G}$ is generated by the Borel subsets of $[0, a]$.

6

Discrete-time martingales

Fix a probability space $(\Omega, \mathcal{F}, P)$. Observation of the values of some random variable $Y : \Omega \to \mathbb{R}$ is often used to estimate ('predict') the behaviour of some unknown variable X. Examples include economic data used to estimate future stock prices and weather data used in forecasting. At its simplest, prediction involves finding some function $f(Y)$ which takes us 'closest' to X; we call $\widehat{X} = f(Y)$ an *estimator* of X. If $\widehat{X}$ minimises the distance from X in the L^2-norm (i.e. $||X - \widehat{X}||_2$ is minimised by $\widehat{X}$ among $\sigma(Y)$-measurable random variables), then statisticians call it the *least-mean-square* estimator of X. In other words, $\widehat{X}$ is the element of $L^2(\sigma(Y))$ closest to $X \in L^2(\mathcal{F})$, i.e. the orthogonal projection of X onto this complete subspace. So the best predictor of X given Y (hence given $\sigma(Y)$) is simply the conditional expectation $\mathbb{E}[X|\sigma(Y)]$.

This suggests, more generally, that we can regard any sub-σ-field $\mathcal{G}$ of $\mathcal{F}$ as providing *partial information* about a random variable $X \in L^1(\mathcal{F})$ and that our *best estimate* of X, given $\mathcal{G}$, is a version of $\mathbb{E}[X|\mathcal{G}]$, which was defined in Chapter 5 when $X \in L^2$. Theorem 6.36 will complete the picture for $X \in L^1$.

6.1 Discrete filtrations and martingales

If X is itself $\mathcal{G}$-measurable, we have $\mathbb{E}[X|\mathcal{G}] = X$ a.s.(P) and our best estimate is perfect up to P-null sets.

Example 6.1 At the other extreme, suppose that $(Y_k)_{k \geq 1}$ is a sequence of independent integrable random variables with $\mathbb{E}[Y_k] = 0$ for all k, and $X_0 = 0$, $X_n = Y_1 + Y_2 + \ldots + Y_n$. We set $\mathcal{F}_0 = \{\emptyset, \Omega)\}$, $\mathcal{F}_n = \sigma(Y_1, Y_2, \ldots, Y_n)$ for $n \geq 1$, so that $\mathcal{F}_n$ represents the 'history' of the (discrete-time) stochastic process $(Y_k)_{k \geq 1}$ up to time n. (For example, the Y_k could be daily fluctuations of a stock price that starts at some known value on day 0.) The best predictor of X_n given $\mathcal{F}_{n-1}$ is given for $n > 0$ by

$$\mathbb{E}[X_n | \mathcal{F}_{n-1}] = \mathbb{E}[X_{n-1} | \mathcal{F}_{n-1}] + \mathbb{E}[Y_n | \mathcal{F}_{n-1}]$$
$$= X_{n-1} + \mathbb{E}[Y_n] = X_{n-1} \text{ a.s } (P),$$

as $\sigma(Y_n)$ and $\mathcal{F}_{n-1}$ are independent, and $X_{n-1} \in L^1(\mathcal{F}_{n-1})$. Thus our best estimate of X_n at time $(n-1)$ is simply the current value X_{n-1}. This expresses the intuitive idea that if the Y_i are independent, then we have no better means of predicting future behaviour of their sums than to work with our current knowledge of X_{n-1}, i.e. the sums (X_n) change in a 'completely random' manner as n increases.

We capture these ideas in the next definition:

Definition 6.2 (i) An increasing sequence $\mathbb{F} = (\mathcal{F}_n)_{n \geq 0}$ of sub-σ-fields of $\mathcal{F}$ is a *filtration*. A sequence $(X_n)_{n \geq 0}$ of random variables is *adapted* to $\mathbb{F}$ if for each n, X_n is $\mathcal{F}_n$-measurable. We call $(\Omega, \mathcal{F}, \mathbb{F}, P)$ a *filtered space*.

(ii) An adapted sequence $(X_n, \mathcal{F}_n)_n$ is an $(\mathbb{F}, P)$-*martingale* (often simply a *martingale*) if each X_n is integrable and

$$\mathbb{E}[X_n | \mathcal{F}_{n-1}] = X_{n-1} \text{ a.s.}(P) \text{ for each } n \geq 1.$$

If $=$ above is replaced by $\leq$, we call $(X_n, \mathcal{F}_n)_n$ a *supermartingale;* if we have $\geq$ instead, it is a *submartingale*. If the X_n are in L^p for $p \geq 1$, we call $(X_n, \mathcal{F}_n)_n$ an L^p-*martingale*. We write X (or $(X, \mathbb{F})$) for $(X_n, \mathcal{F}_n)_n$ when the context is clear.

Remark 6.3 Clearly, X is a submartingale iff $-X$ is a supermartingale, and is a martingale iff it is both a sub- and a supermartingale. Also, changing the random variable X_0 in the sequence $(X_n)_{n \geq 0}$ does not change these properties: the sequence is martingale (resp. super-, sub-) iff the same holds for $(X_n - X_0)_{n \geq 0}$. So we frequently assume $X_0 = 0$ a.s.(P) without loss. The set of $(\mathbb{F}, P)$-martingales is a vector space by Proposition 5.23(i).

Remark 6.4 The properties of conditional expectation in Proposition 5.23 imply many important facts about martingales. By the tower property (5.23(vii)), $\mathbb{E}[X_m|\mathcal{F}_n] = X_n$ a.s.(P) when $m > n$, and similarly for super- and submartingales. Also, by 5.23(ii) the expectation of a martingale is constant ($\mathbb{E}[X_m] = \mathbb{E}(\mathbb{E}[X_m|\mathcal{F}_n]) = \mathbb{E}[X_n]$), while the expectations of supermartingales decrease and those of submartingales increase as n increases. The Jensen inequality, applied to $g(x) = x^2$ shows that when $(X_n, \mathcal{F}_n)_n$ is an L^2-martingale then $\left(X_n^2, \mathcal{F}_n\right)_n$ is a submartingale.

Exercise 6.5 Suppose that $(X_n, \mathcal{F}_n)_n$ is a martingale and $\mathbb{G} = (\mathcal{G}_n)_{n \geq 0}$ with $\mathcal{G}_0 = \{\emptyset, \Omega\}$, $\mathcal{G}_n = \sigma(X_1, \ldots, X_n)$ for each $n \geq 1$. Show that $(X_n, \mathcal{G}_n)_n$ is a martingale. (*Hint:* first verify that $\mathbb{G}$ is the smallest filtration to which $(X_n)_n$ is adapted.)

The partial sums $(X_n)_{n \geq 1}$ of Example 6.1 form a martingale on the filtered space $(\Omega, \mathcal{F}, \mathbb{F}, P)$ with $\mathcal{F}_n = \sigma(Y_1, \ldots, Y_n)$ and $\mathbb{F} = (\mathcal{F}_n)_{n \geq 1}$. At each stage, the σ-field $\mathcal{F}_n$ represents what is known about the $(Y_i)_{i \leq n}$, and this 'information' increases with n. 'Full' information about the random behaviour in question resides in the larger σ-field $\mathcal{F}$, and we shall frequently assume that $\mathcal{F} = \mathcal{F}_\infty$ where we define $\mathcal{F}_\infty = \sigma(\cup_{n \geq 1}\mathcal{F}_n)$ – note that, unlike the intersection, a union of σ-fields need *not* be a σ-field!

Example 6.6 An example where the 'limiting' random variable in a martingale seems 'given' is the following: fix a filtered space $(\Omega, \mathcal{F}, \mathbb{F}, P)$ and $X \in L^1(\Omega, \mathcal{F}, P)$. Let $M_n = \mathbb{E}[X|\mathcal{F}_n]$ for each n. Then M_n is our best estimate of X given the information contained in $\mathcal{F}_n$. Again by the tower property $(M_n, \mathcal{F}_n)$ is a martingale

$$\mathbb{E}[M_n|\mathcal{F}_{n-1}] = \mathbb{E}[\mathbb{E}[X|\mathcal{F}_n]|\mathcal{F}_{n-1}] = \mathbb{E}[X|\mathcal{F}_{n-1}] = M_{n-1}.$$

Our 'best ever' estimate of X is $\lim_n E[X|\mathcal{F}_n]$ if this a.s. limit exists. Logically, it should be $E[X|\mathcal{F}_\infty]$ and if $\mathcal{F}_\infty = \mathcal{F}$ we recover 'full knowledge' of X. These limit theorems, however, require proof, and are discussed below.

Martingales represent fair gambling games, as our best guess at time n is that winnings and losses in the next game cancel out on average – as in Example 6.1, assuming $X_0 = 0$, the *martingale difference* $\Delta X_n = X_n - X_{n-1}$ has $\mathbb{E}[\Delta X_n|\mathcal{F}_{n-1}] = 0$ for all $n \geq 1$. (Equality is replaced by $\geq$ for a submartingale, and by $\leq$ for a supermartingale, so these games are favourable to the gambler and the casino respectively!)

Example 6.7 ('Betting the martingale') The name 'martingale' (which is shared with a type of bridle used in horse racing) has a long history in gambling. In a simple coin-tossing game with a fair coin, the *martingale strategy* is to bet \$1 on the first toss and double the stake for the next if we lose, etc. Suppose our *first* win is at the Nth game. Our net winnings are then $2^N - \left(\sum_{k=0}^{N-1} 2^k\right) = 1$. The random variable N has $P(N=k) = \frac{1}{2^k}$ since we win or lose each toss with probability $\frac{1}{2}$, and we have lost $k-1$ times before winning at game k. (We assume the outcomes of tosses are independent.) So $P(N < \infty) = P(\cup_{k \geq 1}(N=k)) = \sum_{k=1}^{\infty} P(N=k) = 1$. Although we are almost sure of winning eventually, we must be willing to risk a very large fortune before this occurs! The total amount X we would have lost before winning has expectation $\mathbb{E}[X] = \sum_{k=1}^{\infty} X_k P(N=k)$, where $X_k = 2^{k-1} - 1$ is the total wagered (and lost!) *before* winning game k. Thus $\mathbb{E}[X] = \lim_{m \to \infty} \left(\sum_{k=1}^{m}(\frac{1}{2} - \frac{1}{2^k})\right) = \infty$. The dangers of this strategy were obvious to gamblers three hundred years ago, and most casinos now refuse bets from players seen to be using such 'doubling strategies'.

Exercise 6.8 Where is the martingale in Example 6.7? Our betting strategy depends at time n on the random outcomes of the previous games, given by Bernoulli random variables Z_i taking each of the values $+1, -1$ with probability $\frac{1}{2}$. All the previous games must result in losses (i.e. $Z_i = -1$) if we are to reach game n. Thus our net winnings are given by the process $S = (S_n)_{n \geq 1}$, $S_n = S_{n-1} + Z_n f(Z_1, Z_2, \ldots, Z_{n-1})$, where f takes the value 2^{n-1} if $Z_i = -1$ for all $i < n$ and 0 otherwise. Show that S is a martingale for the filtration $(\mathcal{F}_n)_{n \geq 1}$, where $\mathcal{F}_n = \sigma\{Z_i : i \leq n\}$.

6.2 The Doob decomposition

Fix a filtered space $(\Omega, \mathcal{F}, \mathbb{F}, P)$ and an adapted sequence $(Y_n)_{n \geq 0}$ (we shall call this a *process* from now on). The difference process ΔY with $\Delta Y_n = Y_n - Y_{n-1}$ can be written for $n \geq 1$ as a sum of two such processes $\Delta M, \Delta A$ via

$$\Delta Y_n = (\Delta Y_n - \mathbb{E}[\Delta Y_n | \mathcal{F}_{n-1}]) + \mathbb{E}[\Delta Y_n | \mathcal{F}_{n-1}] = \Delta M_n + \Delta A_n,$$

where we also define $M_0 = A_0 = 0$ a.s.(P) and then let $M_n = \sum_{k=1}^{n} \Delta M_k$ and $A_n = \sum_{k=1}^{n} \Delta A_k$.

Then we see that:

(i) for $n \geq 1$, $\mathbb{E}[\Delta M_n | \mathcal{F}_{n-1}] = 0$. In other words, $(M, \mathbb{F})$ is a martingale.

(ii) A_n is $\mathcal{F}_{n-1}$-measurable for each $n \geq 1$. This important property deserves a name.

Definition 6.9 We call a process $Z = (Z_n)_{n \geq 1}$ *predictable* if, for each $n \geq 1$, Z_n is measurable with respect to $\mathcal{F}_{n-1}$.

Predictability means that Z_n is 'known' completely by time $n - 1$. This is the opposite of the 'completely random' behaviour of a martingale:

Exercise 6.10 Show that any predictable martingale is a.s. constant.

The adapted process Y has now been decomposed as the sum of a martingale and a predictable process

$$Y_n = Y_0 + M_n + A_n.$$

We call this the *Doob decomposition* of Y. It is *unique*, for if we also have $Y_n = Y_0 + M_n' + A_n'$ for some other martingale M' and predictable process A' with $M_0' = A_0' = 0$, then for each n, $M_n - M_n' = A_n' - A_n$, which shows that $M - M'$ is a predictable martingale beginning at 0, and so stays constant there by Exercise 6.10. We call M the *martingale part* of Y and $A = Y - M$ the *compensator.*

This decomposition takes on particular importance when we begin with a martingale X. In the decomposition $X^2 = X_0^2 + M + A$ of the submartingale X^2 (cf. Remark 6.4), the predictable process A has difference process

$$\Delta A_n = \mathbb{E}\left[X_n^2 | \mathcal{F}_{n-1}\right] - X_{n-1}^2 \geq 0$$

and thus is *increasing*, i.e. $A_n \leq A_{n+1}$ a.s. for every n. If $X_0 = 0$ (as we can assume without loss), we have written $X^2 = M + A$ as the sum of a martingale M and a predictable increasing process A.

We come to a very useful property of L^2-martingales:

Lemma 6.11 *Let X be an L^2-martingale. Note that*

$$(\Delta X_n)^2 = (X_n - X_{n-1})^2 \text{ and } \Delta X_n^2 = \left(X_n^2 - X_{n-1}^2\right).$$

We have

$$\mathbb{E}\left[(\Delta X_n)^2\right] |\mathcal{F}_{n-1}] = \mathbb{E}\left[\Delta X_n^2 |\mathcal{F}_{n-1}\right].$$

Proof $(\Delta X_n)^2 = X_n^2 - 2X_n X_{n-1} + X_{n-1}^2$, so $\mathbb{E}[(\Delta X_n)^2 |\mathcal{F}_{n-1}]$ equals $\mathbb{E}\left[X_n^2|\mathcal{F}_{n-1}\right] - 2X_{n-1}\mathbb{E}[\Delta X_n|\mathcal{F}_{n-1}] + X_{n-1}^2$, which reduces to $\mathbb{E}[(X_n^2 - X_{n-1}^2)]|\mathcal{F}_{n-1}]$ as claimed.

Given a *martingale* X with $X_0 = 0$, the decomposition $X^2 = M + A$ implies $\Delta M_n = \Delta X_n^2 - \Delta A_n$. But M is also a martingale, so the lemma yields

$$0 = \mathbb{E}[\Delta M_n|\mathcal{F}_{n-1}] = \mathbb{E}[(\Delta X_n)^2|\mathcal{F}_{n-1}] - \mathbb{E}[\Delta A_n|\mathcal{F}_{n-1}].$$

Since A is predictable, $\mathbb{E}[(\Delta X_n)^2|\mathcal{F}_{n-1}] = \Delta A_n$, which exhibits the process A as a 'conditional quadratic variation' process of the original martingale X. Taking expectations, $\mathbb{E}[(\Delta X_n)^2] = \mathbb{E}[\Delta A_n]$.

Also, $\mathbb{E}\left[X_n^2\right] = \mathbb{E}[M_n] + \mathbb{E}[A_n] = \mathbb{E}[A_n]$, so both sides are bounded for all n iff the martingale X is bounded in $\mathcal{L}^2(\Omega, \mathcal{F}, P)$. Since (A_n) is increasing, the a.s. limit $A_\infty(\omega) = \lim_{n \to \infty} A_n(\omega)$ exists, and the boundedness of the integrals ensures in that case that $\mathbb{E}[A_\infty] < \infty$.

Exercise 6.12 Suppose $(Z_n)_{n \geq 1}$ is a sequence of Bernoulli random variables, with each Z_n taking the values 1 and -1, with equal probability. Let $X_0 = 0$, $X_n = Z_1 + Z_2 + \cdots + Z_n$ for $n \geq 1$, and let $(\mathcal{F}_n)_n$ be the natural filtration generated by the (Z_n). Show that the compensator (A_n) in the Doob decomposition of the submartingale $\left(X_n^2\right)_n$ is deterministic (i.e. non-random).

6.3 Discrete stochastic integrals

In discrete time, we easily construct 'stochastic integrals' and show that they preserve the martingale property. We wish to use a martingale X as an *integrator*, in the same sense as we form an integral with the distribution function of a random variable. Of course, for discrete distributions the 'integral' is simply an appropriate linear combination of increments of the distribution function. Thus the key components are linear combinations of the *increments* $\Delta X_n = X_n - X_{n-1}$. However, here we deal with stochastic processes (i.e. functions of both n and ω) rather than real functions, so measurability conditions are needed to determine what constitutes an 'appropriate' linear combination. For $\omega \in \Omega$ we set $I_0(\omega) = 0$ and for $n \geq 1$ let

$$I_n(\omega) = \sum_{k=1}^{n} c_k(\omega)(\Delta X_k)(\omega) = \sum_{k=1}^{n} c_k(\omega)(X_k(\omega) - X_{k-1}(\omega)),$$

choosing $(c_n)_n$ to ensure that the new process $(I_n)_n$ has useful properties. When $(c_n)_n$ is a *bounded predictable* process and X is a martingale, both for the same filtration $(\mathcal{F}_n)_n$, the process $(I_n)_n$ is called a *discrete stochastic integral* (also known as a *martingale transform*). We show that $I = (I_n)_n$ is again a martingale

$$\mathbb{E}[I_n|\mathcal{F}_{n-1}] = \mathbb{E}[(I_{n-1} + c_n\Delta X_n)|\mathcal{F}_{n-1}]$$
$$= I_{n-1} + c_n\mathbb{E}[\Delta X_n|\mathcal{F}_{n-1}] = I_{n-1}.$$

So the discrete stochastic integral preserves the martingale property. We write the stochastic integral as $c \cdot X$, meaning that for all $n \geq 0$, $I_n = (c \cdot X)_n$. We record this as a theorem:

Theorem 6.13 *Let $(\Omega, \mathcal{F}, (\mathcal{F}_n)_{n\geq 0}, P)$ be a filtered probability space. If X is a martingale and c is a bounded predictable process, then the discrete stochastic integral $c \cdot X$ is again a martingale.*

We use the *boundedness* assumption in order to ensure that $c_k\Delta X_k$ is integrable, so that its conditional expectation makes sense. For L^2-martingales, the Schwarz inequality ensures we only need $c_n \in L^2(\mathcal{F}_{n-1})$ for each n.

Remark 6.14 All this is bad news for gamblers: the fact that $c \cdot X$ is again a martingale means that 'clever' gambling strategies will be of no avail when the game is fair. It remains fair whatever strategy the gambler employs! If it starts out unfavourable to the gambler, so that X is a supermartingale ($\mathbb{E}[X_n|\mathcal{F}_{n-1}] \leq X_{n-1}$), the above calculation shows that, as long as $c_n \geq 0$ for each n, then $\mathbb{E}[I_n|\mathcal{F}_{n-1}] \leq I_{n-1}$, so that the game remains unfavourable whatever non-negative stakes the gambler places.

Combining the definition of $(I_n)_n$ with the Doob decomposition of the submartingale X^2 we obtain the key identity which illustrates why martingales make useful 'integrators'. We calculate the expected value of the square of $(c \cdot X)_n$ when $c = (c_n)$ with $c_n \in L^2 (\mathcal{F}_{n-1})$ for each n and $X = (X_n)$ is an L^2- martingale

$$\mathbb{E}\left[(c \cdot X)_n^2\right] = \mathbb{E}\left(\left[\sum_{k=1}^{n} c_k \Delta X_k\right]^2\right) = \mathbb{E}\left[\sum_{j,k=1}^{n} c_j c_k \Delta X_j \Delta X_k\right].$$

Consider terms in the double sum separately: if $j < k$

$$\mathbb{E}[c_j c_k \Delta X_j \Delta X_k] = \mathbb{E}[\mathbb{E}[c_j c_k \Delta X_j \Delta X_k | \mathcal{F}_{k-1}]]$$
$$= \mathbb{E}[c_j c_k \Delta X_j \mathbb{E}[\Delta X_k | \mathcal{F}_{k-1}]] = 0$$

since the first three factors are all $\mathcal{F}_{k-1}$- measurable, while

$$\mathbb{E}[\Delta X_k | \mathcal{F}_{k-1}] = 0$$

since X is a martingale. With j, k interchanged this also shows that these terms are 0 when $k < j$. The remaining terms have the form

$$\mathbb{E}\left[c_k^2 (\Delta X_k)^2\right] = \mathbb{E}\left[c_k^2 \mathbb{E}[(\Delta X_k)^2 | \mathcal{F}_{k-1}]\right] = \mathbb{E}\left[c_k^2 \Delta A_k\right].$$

By linearity, we obtain a fundamental identity for discrete stochastic integrals relative to martingales (also called the *Itô isometry*)

$$\mathbb{E}\left[\left(\sum_{k=1}^{n} c_k \Delta X_k\right)^2\right] = \mathbb{E}\left[\sum_{k=1}^{n} c_k^2 \Delta A_k\right].$$

The sum inside the expectation sign on the right is a 'Stieltjes sum' for the increasing process, so that it is now at least plausible that this identity should allow us to define martingale integrals in the continuous-time setting, using approximation of processes by simple processes. The Itô isometry is of critical importance in the definition of stochastic integrals relative to processes such as Brownian Motion, as we shall see.

6.4 Doob's inequalities

J.L. Doob proved two simple but powerful inequalities for submartingales that explore the relationship between a process and its maximum values up to a fixed time. They will be used frequently in the sequel. Given any process $M = (M_n)_{n \geq 0}$ on $(\Omega, \mathcal{F}, P)$, we define $M_k^* = \max_{n \leq k} M_n$ for each $k \geq 0$.

Theorem 6.15 *(Doob's submartingale inequality)*
If M is a non-negative submartingale and $\lambda > 0$, then for all k

$$\lambda P\left(M_k^* \geq \lambda\right) \leq \int_{\{M_k^* \geq \lambda\}} M_k dP \leq \mathbb{E}[M_k].$$

Proof Let $A = \{M_k^* \geq \lambda\}$, then $A = \cup_{n=0}^k A_n$, where

$$A_0 = \{M_0 \geq \lambda\}, A_n = \left(\cap_{m=0}^{n-1}\{M_m < \lambda\}\right) \cap \{M_n \geq \lambda\}$$

and the union is disjoint. For $0 \leq n \leq k$, $A_n \in \mathcal{F}_n$, so as M is a submartingale and $M_n > \lambda$ on A_n, $\int_{A_n} M_k dP \geq \int_{A_n} M_n dP \geq \lambda P(A_n)$. Now sum over $n = 0, 1, \ldots, k$ to obtain the first inequality; the second is trivial since $M_k \geq 0$ a.s.(P).

Corollary 6.16 *For any $p > 1$, $\lambda^p P\left(M_k^* \geq \lambda\right) \leq \mathbb{E}\left[M_k^p\right]$.*

For the proof, simply apply Theorem 6.15 to the submartingale M_n^p – recall Jensen's inequality!

We combine Theorem 6.15 with Proposition 5.12 to compare the L^p-norms of the sequences (M_n) and (M_k^*).

Theorem 6.17 *(Doob's L^p inequality)*
Suppose that $p > 1$ and $\frac{1}{p} + \frac{1}{q} = 1$. If M is a non-negative submartingale, then $\|M_n^*\|_p \leq q \|M_n\|_p$ for all $n \geq 0$. Hence if M is L^p-bounded, we have $\left\|\sup_{n \geq 0} M_n\right\|_p \leq q \sup_{n \geq 0} \|M_n\|_p$.

Proof Fix $n \geq 0$. Theorem 6.15 allows us to apply Proposition 5.12 with $X = M_n, Y = M_n^*$, so that $\|M_n^*\|_p \leq q \|M_n\|_p \leq q \sup_n \|M_n\|_p$ for each $n \geq 0$. But $M_n^* \uparrow \sup_n M_n$, so by the MCT the second claim follows at once.

6.5 Martingale convergence

Recall Example 6.6: when $M_n = \mathbb{E}[X|\mathcal{F}_n]$ for some fixed X in $L^1(\Omega, \mathcal{F}, P)$ and filtration $\mathbb{F} = (\mathcal{F}_n)_{n \geq 0}$, it is natural to ask whether $M_n \to X$ in some sense when $n \to \infty$. The two modes of convergence of interest are $M_n \to X$ a.s.(P) and $\|M_n - X\|_p \to 0$, for $p \geq 1$, i.e. *almost sure* and convergence in L^p-*norm*. We give only a brief overview of this theory. More detail can be found in [BZ],[K], [N] or [W].

If a real sequence $(x_n)_n$ *diverges*, $\liminf_n x_n < \limsup_n x_n$. As $\mathbb{Q}$ is dense in $\mathbb{R}$, we can find rationals $a < b$ such that $\liminf_n x_n < a < b < \limsup_n x_n$. So for a sequence $(X_n)_n$ of random variables we can express the set of $\omega \in \Omega$, where $(X_n(\omega))_n$ has no limit as a countable union of sets $D = \cup_{\substack{a<b \\ a,b \in \mathbb{Q}}} D_{a,b}$, where

$$D_{a,b} = \{\omega : \liminf_n X_n(\omega) < a < b < \limsup_n X_n(\omega)\}.$$

The task of proving a.s. (P)-convergence of (X_n) now consists of showing that each $D_{a,b}$ is P-null. For supermartingales *bounded in* L^1, i.e., such that $\sup_n \|X_n\|_1 < \infty$, this is guaranteed by the famous Doob Upcrossing Lemma. We need some notation.

Definition 6.18 Let $(X_n, \mathcal{F}_n)_{n \geq 0}$ be adapted and let a, b be real numbers with $a < b$. Define a gambling strategy $c = (c_n)_{n \geq 1}$ for (X_n) and $[a, b]$ by setting $c_1 = 0$ and, for each $n \geq 1$, let $c_{n+1} = 1$ if either $c_n = 0$ *and* $X_n < a$, or $c_n = 1$ *and* $X_n \leq b$. Otherwise, set $c_{n+1} = 0$. We see inductively that c is predictable.

Then define an *upcrossing* of $[a, b]$ as an integer u such that $c_u = 1$ and $c_{u+1} = 0$. The *number* $U_N[a, b]$ *of upcrossings* of $[a, b]$ by $n \to X_n(\omega)$ by time N is then the largest k such that the upcrossings satisfy $u_0 = 0 < u_1 < u_2 < \ldots u_k \leq N$.

The strategy $c = (c_n)_{n \geq 1}$ represents a peculiar gambling strategy for a game where $\Delta X_n = X_n - X_{n-1}$ denotes the winnings per unit stake in each round. We do not enter the game until $X_n < a$, then bet a unit stake in each round until $X_n > b$, and wait until $X_n < a$ before making the next unit bet, and so on. This is a bounded predictable process, so Theorem 6.13 applies to $Y = c \cdot X$. The process $Y = c \cdot X$ describes our winnings using this strategy.

Exercise 6.19 Figure 6.1 shows an example with three upcrossings of $[a, b]$, marked by bold line segments. Sketch the dynamics of the process Y corresponding to the sketched process X.

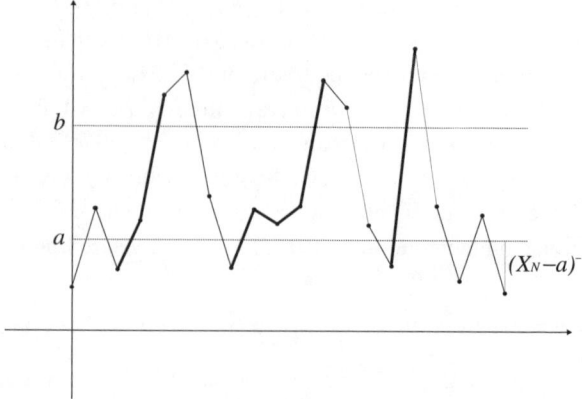

Fig. 6.1 Upcrossings of $[a, b]$ by X

Lemma 6.20 *(Doob's Upcrossing Lemma) If (X_n) is a supermartingale and $a < b$, then for any N*

$$(b - a)\mathbb{E}[U_N[a, b]] \leq \mathbb{E}[(X_N - a)^-].$$

Proof $Y_0 = 0$ a.s. by definition and $Y_N = \sum_{i=1}^N c_i \Delta X_i$. The number of upcrossings is $U_N[a, b] = k$, where $u_k \leq N < u_{k+1}$ by definition. Each upcrossing increases Y by at least $b - a$, so we have $Y_{u_i}(\omega) - Y_{u_{i-1}}(\omega) \geq b - a$ for $i \leq k$. Moreover, we have $Y_N(\omega) - Y_{u_k}(\omega) \geq -(X_N - a)^-$, since the 'loss' in the periods of play after the last upcrossing cannot exceed $(X_N - a)^-$. Hence $Y_N(\omega) \geq (b - a)U_N[a, b] - (X_N - a)^-$. By Theorem 6.13, Y is a supermartingale null at 0, so $\mathbb{E}[Y_N] \leq 0$.

For $p \geq 1$, a family $\mathcal{S}$ in L^p is L^p-*bounded* if there is a B with $\|X\|_p < B$ for all X in $\mathcal{S}$. Our first convergence theorem applies to L^1-bounded supermartingales. This includes non-negative supermartingales:

Exercise 6.21 Show that if the supermartingale X has $X_n \geq 0$ a.s. for all n, then X is L^1-bounded.

Theorem 6.22 *Let $(X_n)_n$ be an L^1-bounded supermartingale. Then $(X_n)_n$ converges a.s. (P) to the $\mathcal{F}_\infty$-measurable random variable $X_\infty = \limsup_n X_n$.*

Proof We show that $U_N[a, b]$ is a.s. finite for each N. By Lemma 6.20

$$(b - a)\mathbb{E}[U_N[a, b]] \leq \mathbb{E}[|X_N|] + |a| \leq \sup_n ||X_n||_1 + |a|.$$

Hence $P(U_N[a, b] = \infty) = 0$ for all N. This proves the a.s. convergence. The limit X_∞ is a.s. finite: by Fatou

$$\mathbb{E}[|X_\infty|] = \mathbb{E}[\liminf_n |X_n|] \leq \liminf_n \mathbb{E}[|X_n|]$$
$$\leq \limsup_n \mathbb{E}[|X_n|] < \infty.$$

However, L^1-boundedness of a supermartingale $(X_n)_n$ is not sufficient for convergence to X_∞ in L^1-norm. To see what is needed, note that if $Z \in L^1$, then for any given $\varepsilon > 0$ we can find $K > 0$ such that $\int_{\{|Z|>K\}} |Z| \, dP < \varepsilon$: take $Z \geq 0$ and write $\mathbb{E}[Z] = \int_{\{Z \leq K\}} Z dP + \int_{\{Z > K\}} Z dP$. Now $Z_n = Z \mathbf{1}_{\{Z \leq n\}} \uparrow Z$ as $n \uparrow \infty$, hence by the MCT $\mathbb{E}[Z_n] \uparrow \mathbb{E}[Z]$ and so, for given $\varepsilon > 0$, by taking n large enough we have $\mathbb{E}[Z] - \int_{\{Z \leq n\}} Z dP < \varepsilon$. Take any $K > n$.

Exercise 6.23 Prove, conversely, that if given $\varepsilon > 0$, there is $K > 0$ such that $\int_{\{|Z|>K\}} |Z| dP < \varepsilon$, then $Z \in L^1$. Thus Z is integrable iff the above condition holds.

We now extend the condition to families (e.g. sequences) of random variables.

Definition 6.24 $S \subset L^1(\Omega)$ is *uniformly integrable (u.i.)* if given $\varepsilon > 0$ there is $K > 0$ such that $\int_{\{|X|>K\}} |X| \, dP < \varepsilon$ for every X in S. (Equivalently: $\sup_{X \in S} \int_{\{|X|>K\}} |X| \, dP \to 0$ as $K \to \infty$.)

Exercise 6.25 Show that if a family of random variables is L^p-bounded for some $p > 1$, then it is *u.i.* (*Hint:* $y > K > 0$ implies $y < \frac{y^p}{K^{p-1}}$. Show that $\int_{\{|X|>K\}} |X| dP < \frac{1}{K^{p-1}} \int_{\{|X|>K\}} |X|^p dP$.)

Example 6.26 $S = \{\mathbb{E}[X|\mathcal{G}]: \mathcal{G}$ is a sub-σ-field of $\mathcal{F}\}$ is uniformly integrable when $X \in L^1$. To see this, fix $K > 0$ and set

$A_K = \{\mathbb{E}[X|\mathcal{G}] > K\} \in \mathcal{G}$. We have $|\mathbb{E}[X|\mathcal{G}]| \leq \mathbb{E}[|X||\mathcal{G}]$ a.s. by Jensen, so $KP(A_K) \leq \int_{A_K} |\mathbb{E}[X|\mathcal{G}]|dP \leq \int_{A_K} |X| dP \leq ||X||_1$. This means that $P(A_K)$, and therefore $\int_{A_K} |X| dP$, goes to 0 *independently of the choice of $\mathcal{G} \subset \mathcal{F}$ as $K \uparrow \infty$.*

Example 6.27 In particular, given $X \in L^1$, the martingale $(M_n)_n$ with $M_n = \mathbb{E}[X|\mathcal{F}_n]$ for $n \geq 0$ is u.i. We show below that *all* u.i. martingales have this form. In preparation, note that any u.i. sequence $(Z_n)_n$ in L^1 is L^1-bounded, i.e. $\sup_n ||Z_n||_1 < \infty$. To see this, choose $K > 0$ with $\int_{\{|Z_n|>K\}} |Z_n| dP < 1$. Then

$$\mathbb{E}[|Z_n|] = \int_{\{|Z_n|\leq K\}} |Z_n| dP + \int_{\{|Z_n|>K\}} |Z_n| dP \leq K + 1.$$

Thus a u.i. supermartingale $(X_n)_n$ converges a.s. to X_∞.

To arrive at convergence in L^p-norm, we need another mode of convergence which is useful in many contexts:

Definition 6.28 $(Y_n)_n$ converges to Y *in probability* if $\lim_n P(|Y_n - Y| \geq \varepsilon) = 0$ for all $\varepsilon > 0$.

The next exercise shows that this convergence in weaker than L^p-norm or a.s. convergence:

Exercise 6.29
(i) Show that if $Y_n \to Y$ in L^p-norm or a.s. (P), then $Y_n \to Y$ in probability. (*Hint*: Assume the limit is 0. For the first claim, use Chebychev's inequality.)
(ii) Show that if $Y_n \to Y$ in probability, then some subsequence converges to Y a.s.
(iii) Use the sequence $Y_1 = \mathbf{1}_{[0,1]}, Y_2 = \mathbf{1}_{[0,\frac{1}{2}]}, Y_3 = \mathbf{1}_{[\frac{1}{2},1]}$, $Y_4 = \mathbf{1}_{[0,\frac{1}{4}]}, \dots$ to show that convergence in probability does not imply a.s. convergence.

However, convergence in probability implies L^1-convergence for u.i. sequences:

Proposition 6.30 *Let $(Y_n)_n$ be a u.i. L^1-sequence such that $Y_n \to 0$ in probability. Then $||Y_n||_1 \to 0$ as $n \to \infty$.*

Proof As the sequence is u.i., given $\varepsilon > 0$ we can find $K > \frac{\varepsilon}{3}$ such that $\int_{\{|Y_n|>K\}} |Y_n| \, dP < \frac{\varepsilon}{3}$. But $\lim_n P(|Y_n| > \varepsilon) = 0$ for every $\varepsilon > 0$, as $Y_n \to 0$ in probability. So, given $\varepsilon > 0$, find $N > 0$ with $P(|Y_n| > \frac{\varepsilon}{3}) < \frac{\varepsilon}{3K}$ if $n \geq N$. For $n \geq 1$, set

$$A_n = \{|Y_n| > K\}, B_n = \left\{K \geq |Y_n| > \frac{\varepsilon}{3}\right\}, C_n = \left\{|Y_n| \leq \frac{\varepsilon}{3}\right\}.$$

$$\mathbb{E}[|Y_n|] = \int_{A_n} |Y_n| \, dP + \int_{B_n} |Y_n| \, dP + \int_{C_n} |Y_n| \, dP$$
$$\leq \frac{\varepsilon}{3} + KP\left(|Y_n| > \frac{\varepsilon}{3}\right) + \frac{\varepsilon}{3} P\left(|Y_n| \leq \frac{\varepsilon}{3}\right).$$

Hence $\|Y_n\|_1 \to 0$ as $n \to \infty$.

Exercise 6.31 Show that the converse of this proposition also holds, so that these two conditions characterise L^1-convergence. (See [W], Ch. 13 if you get stuck.)

Theorem 6.32 *Every u.i. supermartingale $(X_n)_{n \geq 0}$ converges a.s.(P) and in L^1-norm to a limit $X_\infty \in L^1$.*

Proof By Theorem 6.22, $X_n \to X_\infty$ a.s., and using $Y_n = X_n - X_\infty$ instead we have $Y_n \to 0$ a.s.(P). This means that $Y_n \to 0$ in probability, and is u.i., so Proposition 6.30 yields $\|Y_n\|_1 \to 0$.

Corollary 6.33 *Every u.i. $(\mathcal{F}_n)$-martingale $M = (M_n)_{n \geq 0}$ has the form $M_n = \mathbb{E}[Y|\mathcal{F}_n]$ for some $Y \in L^1$ and all n.*

Proof Let $k > n$, then by the martingale property, for any $A \in \mathcal{F}_n$, we have $\int_A M_k dP = \int_A M_n dP$. So with $Y = M_\infty$, the a.s. limit of $(M_n)_n$, we obtain for such A and $k > n$

$$\left| \int_A (M_k - Y) dP \right| \leq \int_A |M_k - Y| \, dP \leq \|M_k - Y\|_1.$$

By Theorem 6.32, $\|M_k - Y\|_1 \to 0$ as $k \to \infty$, so $\int_A M_n dP = \int_A Y dP$ for all $A \in \mathcal{F}_n$. Thus $M_n = \mathbb{E}[Y|\mathcal{F}_n]$ for each n.

Remark 6.34 Given X in L^1 and a filtration $\mathbb{F} = (\mathcal{F}_n)_{n \geq 0}$, we can construct a u.i. martingale by setting $M_n = \mathbb{E}[X|\mathcal{F}_n]$. The a.s. limit

M_∞ of this sequence can now be identified as $M_\infty = \mathbb{E}[X|\mathcal{F}_\infty]$, where $\mathcal{F}_\infty = \sigma(\cup_n \mathcal{F}_n)$ as in Example 6.6. To do this, assume $X \geq 0$ without loss. Now $P_1(A) = \int_A M_\infty dP$ and $P_2(A) = \int_A \mathbb{E}[X|\mathcal{F}_\infty]dP$ are measures on $(\Omega, \mathcal{F}_\infty)$ and they agree on every $A \in \mathcal{F}_n$ (check this carefully!). Since $\cup_{n \geq 0} \mathcal{F}_n$ is a field, it is a π-system, so these measures agree on $\mathcal{F}_\infty$. Both $M_\infty = \limsup_n M_n$ and $\mathbb{E}[X|\mathcal{F}_\infty]$ are $\mathcal{F}_\infty$-measurable, hence so is the set $\{\mathbb{E}[X|\mathcal{F}_\infty] > M_\infty\} = A$. But $P_1(A) = P_2(A)$, so $\int_A (\mathbb{E}[X|\mathcal{F}_\infty] - M_\infty)dP = 0$. Thus $P(A) = 0$ and similarly with roles reversed, so that $M_\infty = \mathbb{E}[X|\mathcal{F}_\infty]$ a.s.(P).

6.6 The Radon–Nikodym Theorem

As an application of Theorem 6.32, we prove the key structure theorem in measure theory, which also shows that conditional expectations exist for any integrable random variables, with properties as proved in Proposition 5.23 and extending our earlier construction for L^2-functions.

Definition 6.35 Suppose P, Q are probabilities on $(\Omega, \mathcal{F})$ and that $P(F) = 0$ implies $Q(F) = 0$ when $F \in \mathcal{F}$. We write $Q \ll P$ and say that Q is *absolutely continuous* with respect to P.

This implies (and is in fact equivalent to) the following *continuity* condition: given $\varepsilon > 0$, we can find $\delta > 0$ such that $P(F) < \delta$ implies $Q(F) < \varepsilon$. (Sketch of proof: if this condition fails, we can find $\varepsilon > 0$ and $(A_n)_n$ with $P(A_n) < \frac{1}{2^n}$ for each n but $Q(A_n) \geq \varepsilon$. But then $P(\limsup_n A_n) = 0 < \varepsilon \leq Q(\limsup_n A_n.)$

When $Q \ll P$ we want to demonstrate the existence of $X \in L^1$ such that for $F \in \mathcal{F}$, $Q(F) = \int_F X dP$. We then write $X = \frac{dQ}{dP}$ and call it the *Radon–Nikodym derivative* of Q w.r.t. P. (Note that Proposition 4.19 tells us that, conversely, Q defined via X in this way is a measure, and obviously $Q \ll P$.)

We construct X in a special case. If the σ-field $\mathcal{G}$ is generated by a sequence $(G_n)_n$ of sets in $\mathcal{F}$, we call it *separable*. Let $\mathcal{G}_n = \sigma(G_1, \ldots, G_n)$ for each $n \geq 1$. These σ-fields are generated by finite partitions of Ω (dividing Ω into *atoms*, i.e. finite intersections $\cap_{i \leq n} F_i$, where each F_i is either G_i or G_i^c, so that no non-empty proper subset of this intersection belongs to $\mathcal{G}_n$). Let $\mathcal{P}_n$ be the smallest partition generating $\mathcal{G}_n$; clearly, $\mathcal{P}_{n+1}$ refines $\mathcal{P}_n$, so each set in $\mathcal{P}_n$ is a *disjoint union* of sets in $\mathcal{P}_{n+1}$. Define $X_n = \sum_{A \in \mathcal{P}_n} \frac{Q(A)}{P(A)} \mathbf{1}_A$ if $P(A) > 0$ and 0 otherwise. To see that (X_n) is a martingale for $\mathbb{G} = (\mathcal{G}_n)_n$, we note that for $A \in \mathcal{P}_n$

$$\int_A X_{n+1} dP = \sum_{B \in \mathcal{P}_{n+1}, B \subset A} \frac{Q(B)}{P(B)} P(B) = Q(A) = \int_A X_n dP.$$

(If $P(B) = 0$, then $Q(B) = 0$, so the finite sum counts all $B \subset A$ with $Q(B) > 0$.) Moreover, $(X_n)_n$ is u.i.: given $\varepsilon > 0$ choose $\delta > 0$ by the continuity condition and $K > \frac{1}{\delta}$. Then with $A = (X_n > K)$, we have $KP(A) \leq ||X_n||_1 = Q(\Omega) < K\delta$. Hence $\int_A X_n dP = Q(A) < \varepsilon$.

Thus $X_n \to X_\infty$ a.s. and in L^1-norm, by Theorem 6.32, and $X_n = \mathbb{E}[X_\infty | \mathcal{G}_n]$ for all n. So for all A in the π-system $\cup_n \mathcal{G}_n$, $\int_A X_\infty dP = \int_A X_k dP = Q(A)$, and the measures Q and $A \mapsto \int_A X_\infty dP$ agree, hence they agree on $\mathcal{G}$.

The final step is to remove the separability assumption. This can be done in various ways (see [K], [W] for two different proofs), but will be omitted here as it really belongs to functional analysis. Thus we have:

Theorem 6.36 *(Radon–Nikodym) Let P, Q be probability measures on $(\Omega, \mathcal{F})$. When $Q \ll P$, we can find a unique $X \in L^1$ such that for $F \in \mathcal{F}$, $Q(F) = \int_F X dP$.*

Corollary 6.37 *If $\mathcal{G}$ is a sub-σ-field of $\mathcal{F}$ and $X \in L^1(\mathcal{F})$, then there is a unique $Y \in L^1(\mathcal{G})$ such that $\int_G X dP = \int_G Y dP$ for all A in $\mathcal{G}$.*

Proof Assume $X \geq 0$ without loss. $Q_X(A) = \int_A X dP$ is absolutely continuous w.r.t P, hence the same holds for their restrictions to $\mathcal{G}$. So the Radon–Nikodym derivative $Y = \frac{dQ_{X|\mathcal{G}}}{dP_\mathcal{G}}$ satisfies, for $G \in \mathcal{G}$, $\int_G X dP = Q_X(G) = Q_{X|\mathcal{G}}(G) = \int_G Y dP$.

Defining $\mathbb{E}[X|\mathcal{G}] = Y$ a.s. (P), this coincides with the orthogonal projection of X onto $L^2(\mathcal{G})$ when $X \in L^2(\mathcal{F})$. We have thus extended the definition of conditional expectation given in Chapter 5 to all integrable random variables. All the properties derived there extend to this setting, with the same proofs.

Exercise 6.38 Radon–Nikodym derivatives obey a simple and natural 'arithmetic', which will be familiar from integration techniques in elementary calculus. Prove the following:

(1) Given probabilities $Q_1 \ll P$ and $Q_2 \ll P$ and setting $Q = Q_1 + Q_2$, we have $\frac{dQ}{dP} = \frac{dQ_1}{dP} + \frac{dQ_2}{dP}$.

(2) If, further, $Q_2 \ll Q_1$, then $\frac{dQ_2}{dP} = \frac{dQ_2}{dQ_1} \frac{dQ_1}{dP}$.

Remark 6.39 We call finite measures P and Q *equivalent* (written $P \sim Q$) if they have the same null sets. In other words, $Q \ll P$ and $P \ll Q$ both hold. From (2) above, we conclude that for equivalent measures the Radon–Nikodym derivatives are strictly positive and satisfy:

$$\frac{dP}{dQ} = \left(\frac{dQ}{dP} \right)^{-1}.$$

At the other extreme, two measures P, Q on $(\Omega, \mathcal{F})$ are *mutually singular* if their 'mass' is *concentrated* on disjoint sets: in other words, there are *disjoint* sets E_P, E_Q such that $P(F) = P(E_P \cap F)$ and $Q(F) = Q(E_Q \cap F)$ for every F in $\mathcal{F}$. (So P is concentrated on E_P and Q on E_Q.) We write $Q \perp P$ for this. For any F disjoint from E_P, we have $P(F) = 0$, and similarly for E_Q. This idea, together with the Radon–Nikodym Theorem, provides insight into the structure of the vector space of all (probability) measures on $(\Omega, \mathcal{F})$:

Theorem 6.40 *(Lebesgue Decomposition Theorem) If P, Q are probabilities on $(\Omega, \mathcal{F})$, then Q can be expressed uniquely as a sum of two measures, $Q_a \ll P$ and $Q_s \perp P$. Hence there is a unique $X \in L^1(P)$ such that $Q(F) = \int_F X \, dP + Q_s(F)$ for every F in $\mathcal{F}$.*

Proof Write $R = P + Q$, so that $0 \leq Q \leq R$ and we can find a measurable h with values in $[0, 1]$ such that $Q(F) = \int_F h \, dR$ for all F in $\mathcal{F}$. The event $A = (h < 1)$ and its complement partition Ω and we set $Q_a(F) = Q(A \cap F), Q_s(F) = Q(A^c \cap F)$ for each F. We show that $Q_a \ll P$ and $Q_s \perp P$: for the first note that if $F \subset A$ is P-null, then $Q(F) = \int_F h \, dR = \int_F h \, dQ$. So $\int_F (1 - h) \, dQ = 0$. But $h < 1$ on A, so $Q(F) = 0$, hence also $Q_a(F) = Q(A \cap F) = 0$. For the second, $F \cap A = \emptyset$ means that $F \subset A^c$, $h = 1$ on F. Hence $Q(F) = \int_F h \, dR = P(F) + Q(F)$, i.e. $P(F) = 0$. Thus P is concentrated on A and Q_s on A^c, which shows that $Q_s \perp P$. The uniqueness claim follows by Exercise 6.41(iii) below, while the final statement is clear from the Radon–Nikodym theorem.

Exercise 6.41 Verify the following claims for probabilities P, Q_1, Q_2 defined on $(\Omega, \mathcal{F})$:
(i) if $Q_i \perp P$ for $i = 1, 2$, then also $(Q_1 + Q_2) \perp P$.
(ii) if $Q_1 \ll P$ and $Q_2 \perp P$, then $Q_1 \perp Q_2$.
(iii) if $Q_1 \ll P$ and $Q_1 \perp P$, then $Q_1 = 0$.

Exercise 6.42 Let random variables X, Y have densities f_X, f_Y. Describe under what conditions on the densities we have $P_X \ll P_Y$ and find the Radon–Nikodym derivative.

7

Brownian Motion

Stochastic processes become an especially powerful modelling tool when the time set is $\mathbb{T} = [0, \infty)$ since we intuitively experience time as a continuous phenomenon. At its simplest, a (*stochastic*) *process* X is a family $(X_t : t \geq 0)$ of random variables defined on the same probability space $(\Omega, \mathcal{F}, P)$. But it is instructive to look at X from several different angles.

7.1 Processes, paths and martingales

Recall the metric space $L^0(\mathcal{F})$ of (equivalence classes of) all random variables $\Omega \rightarrow \mathbb{R}$. A stochastic process X can be viewed as a map $t \rightarrow X_t$ from $\mathbb{T}$ to $L^0(\mathcal{F})$ provided we *identify* X with any of its versions Y, in the following sense:

Definition 7.1 Processes X and Y are *versions* (or *modifications*) of each other if $P(\{\omega \in \Omega : X_t(\omega) = Y_t(\omega)\}) = 1$ for each $t \geq 0$.

So for each $t \geq 0$, the random variables X_t and Y_t are versions of each other as defined in Chapter 5. Unlike in the discrete case, this does not always ensure that the *paths* $t \rightarrow X_t(\omega)$ and $t \rightarrow Y_t(\omega)$ traced out by the processes X, Y will coincide for almost all $\omega \in \Omega$. If $E_t = P(X_t \neq Y_t) = 0$ for each t, the set $E = \cup_{t \geq 0} E_t$ where X and Y differ need no longer be P-null, as the time set is uncountable. We need a stronger condition.

Definition 7.2 Processes X and Y are *indistinguishable* if almost all their paths coincide, i.e. if

$$P(\{\omega \in \Omega : X_t(\omega) = Y_t(\omega) \text{ for all } t \geq 0\}) = 1.$$

Indistinguishable processes are clearly versions of each other, but not conversely. However, matters simplify greatly when almost all paths are continuous (we then say X, Y are continuous processes).

Exercise 7.3 Show that the above two definitions coincide when X and Y are continuous processes. (*Hint*: the positive rationals are dense in $[0, \infty)$.)

As in Chapter 6, the filtration $(\mathcal{G}_t)_{t \geq 0}$ generated by X, where $\mathcal{G}_t = \sigma(X_s : s \leq t)$, models information about the history of the process $X = (X_t)_{t \geq 0}$. More generally, a family $\mathbb{F} = (\mathcal{F}_t)_{t \geq 0}$ of sub-σ-fields of $\mathcal{F}$ is a *filtration* if it is increasing ($\mathcal{F}_s \subset \mathcal{F}_t \subset \mathcal{F}$ for $s \leq t$ in $\mathbb{T}$), and satisfies two technical conditions (the 'usual' conditions):
(i) $\mathcal{F}_0$ contains all P-null sets.
(ii) $\mathbb{F}$ is *right-continuous*, in that $\mathcal{F}_t = \cap_{s>t}\mathcal{F}_s$ for each t in $\mathbb{T}$.
   The family $(\mathcal{G}_t)_{t \geq 0}$ describing the history of a process X will thus be augmented by adding the collection $\mathcal{N}$ of all subsets of P-null sets in $\mathcal{G}_\infty$ (setting $P(N) = 0$ for each $N \in \mathcal{N}$) and defining $\mathcal{F}_t = \sigma(\mathcal{G}_t \cup \mathcal{N})$ for each $t \geq 0$. The *augmented filtration* $(\mathcal{F}_t)_t$ ensures that no 'extra' null sets appear suddenly as the history of the process evolves, and (ii) regulates the occurrence of 'jumps' in our information up to time t, just as for jumps in distribution functions.

Definition 7.4 The process X is *adapted* to a given filtration $\mathbb{F}$ if X_t is $\mathcal{F}_t$-measurable for each t in $\mathbb{T}$.

In particular, for a process $X = (X_t)_{t \geq 0}$, its augmented filtration $(\mathcal{F}_t)_{t \geq 0}$ is the smallest filtration to which X is adapted.
   The stochastic process $X = (X_t)_{t \geq 0}$ can also be viewed as a single map $(t, \omega) \to X_t(\omega)$ from $\mathbb{T} \times \Omega$ to $\mathbb{R}$. This map is *measurable* if $X^{-1}(B)$ is in the product σ-field $\mathcal{B}_\mathbb{T} \times \mathcal{F}$ for every Borel set B in $\mathbb{R}$, where $\mathcal{B}_\mathbb{T}$ consists of the Borel sets contained in $[0, \infty)$. It is *progressive* if for each $t \geq 0$ the map $(s, \omega) \to X_s(\omega)$, defined on $[0, t] \times \Omega$, is $\mathcal{B}_{[0,t]} \times \mathcal{F}_t$-measurable. This is a stronger condition than 'measurable and adapted', but it is not difficult to show (see [K]) that for path-continuous processes the two concepts coincide. It can be shown (though the proof is long and demanding) that every measurable adapted process has a progressive version.

Martingales in continuous time are defined just as their discrete counterparts and have many similar properties:

Definition 7.5 A stochastic process X on $\mathbb{T} \times \Omega$ is an $\mathbb{F}$-*martingale* if $X_t \in L^1(\mathcal{F}_t)$ for all $t \geq 0$ and $\mathbb{E}[X_t|\mathcal{F}_s] = X_s$ for all $s \leq t$.

Remark 7.6 Super- and submartingales are defined by replacing equality by $\leq, \geq$, respectively. The set of $(\mathbb{F}-)$ martingales is a vector space and by Jensen's inequality, if X is a martingale and ϕ a convex real function, then $\phi(X)$ is a submartingale. For any L^2-martingale M and $0 \leq s < t$, we can argue exactly as in Lemma 6.11 to prove the useful identity $\mathbb{E}\left[\left(M_t^2 - M_s^2\right)|\mathcal{F}_s\right] = \mathbb{E}[(M_t - M_s)^2|\mathcal{F}_s]$.

A key feature of martingales are their simple *path properties*. In fact, under the conditions we impose on filtrations it can be shown (see [K]) that any supermartingale has a version with (a.s.) right-continuous paths (and that the limits on the left exist) iff the map $t \rightarrow \mathbb{E}[X_t]$ is right-continuous. But for a martingale this map is a constant, so we can *always* assume that we are working with martingales almost all of whose paths are right-continuous and have left limits. This is usually abbreviated from the French: *càdlàg* (continu à droite et limites à gauche).

We concentrate on *continuous L^2-martingales*. The convergence results for martingales proved in Chapter 6 extend easily, since we can restrict attention to what happens at (dyadic) *rational* indices. Similarly, the continuous-time versions of the two Doob inequalities follow upon applying the discrete-time inequalities to dyadic dissections of the finite interval $[0, T]$. We sketch the proofs briefly.

Proposition 7.7   *Suppose $(M_t)_{t\geq 0}$ is a continuous non-negative submartingale and $\lambda > 0$. If $p \geq 1$, $M_T^* = \sup_{t\in[0,T]} M_t$, then $\lambda P(M_T^* > \lambda) \leq \mathbb{E}[M_T]$, and if $M_T \in L^p(\Omega, \mathcal{F}, P)$ for some $p > 1$, then $\|M_T^*\|_p \leq q \|M_T\|_p$, where $\frac{1}{p} + \frac{1}{q} = 1$.*

Proof   For each n, apply Theorem 6.15 to the finite index sets $D_{n,T} = \left\{\frac{iT}{2^n} : 0 \leq i \leq 2^n\right\}$, so $\lambda P(\sup_{t\in D_{n,T}} M_t > \lambda) \leq \mathbb{E}[M_T]$. Now $\lim_n (\sup_{t\in D_{n,T}} M_t(\omega)) = M_T^*(\omega)$ a.s.(P) by continuity. If $A_n = \{\omega : \sup_{t\in D_{n,T}} M_t(\omega) > \lambda\}$ and $A = \{\omega : M_T^*(\omega) > \lambda\}$, then

$1_{A_n} \uparrow 1_A$ a.s.(P). The MCT now yields: $\lambda P\left(M_T^* > \lambda\right) \leq \mathbb{E}[M_T]$. The second part follows similarly from Theorem 6.17.

The main convergence result for L^p-bounded martingales is:

Theorem 7.8 *If the martingale* $M = (M_t)_{t \geq 0}$ *is* L^p-*bounded for some* $p > 1$, *then there exists* $M_\infty \in L^p$ *such that* $M_t \to M_\infty$ *a.s.*(P) *and in* L^p-*norm as* $t \to \infty$.

Proof   We may assume that M has right-continuous paths a.s.(P). Doob's L^p-inequality, Exercise 6.25 and Theorem 6.32 show that the discrete martingale $(M_n)_{n \geq 0}$ has a.s. limit $M_\infty \in L^p$ and $\|M_n - M_\infty\|_p \to 0$ as $n \to \infty$. For fixed k and real $t \geq k$, $|M_t - M_\infty|$ is bounded above by the sum of $|M_n - M_\infty|$ and $\sup_{t \geq k} |M_t - M_k|$. Thus

$$\limsup_{t \to \infty} |M_t - M_\infty| \leq \lim_{k \to \infty}(\sup_{t \geq k} |M_t - M_k|)$$

since $|M_n - M_\infty| \to 0$. To estimate the RHS, apply Proposition 7.7 to the submartingale $(|M_t - M_k|^p)_{t \geq 0}$ and λ^p instead of M_n and λ to obtain $P(\sup_{t \geq k} |M_t - M_k| > \lambda) \leq \frac{1}{\lambda^p} \mathbb{E}[|M_t - M_k|^p]$.

Letting $t \to \infty$ on the right replaces M_t by M_∞, while letting $k \to \infty$ shows that both quantities converge to 0. By Exercise 1.15, we then have $P(\lim_k \sup_{t \geq k} |M_t - M_k| > \lambda) = 0$ for all $\lambda > 0$, so that $M_t \to M_\infty$ as $t \to \infty$. The L^p-convergence follows since the submartingale $|M_t - M_k|$ is bounded in L^p norm by $|M_n - M_k|$ whenever $n > t$, so

$$\limsup_{t \to \infty} \|M_t - M_\infty\|_p \leq \|M_k - M_\infty\|_p + \sup_{n \geq k} \|M_n - M_k\|_p$$

and both terms on the right go to 0 as $k \to \infty$.

7.2 Convergence of scaled random walks

Recall the symmetric random walk described in Example 6.12: the independent Bernoulli random variables $(X_i)_{i \geq 0}$ can be taken as defined on the 'path' space $\Omega = \{-1, 1\}^{\mathbb{N}}$ (whose elements are functions $\omega : \mathbb{N} \to \{-1, 1\}$) via coordinates: $X_i(\omega) = \omega(i)$. Now $(X_i)_{i \leq n}$ creates a partition $\mathcal{P}_n$ of Ω, which fixes the first n coordinates of each ω in $\mathcal{P}_n$, so that $X_1, \ldots, X_n$ are constant on $\mathcal{P}_n$. Write $\mathcal{F}_n = \sigma(\mathcal{P}_n)$. The symmetric random walk $Y = (Y_n)_n$ with $Y_0 = 0, Y_n = \sum_{i=1}^n X_i$ then has the properties:

(a) Y is centered, i.e. $\mathbb{E}[Y_n] = 0$ for all n.

(b) $Var Y_n = \sum_{i=1}^n Var X_i = n$ (by independence of the (X_i)).

(c) Since for $i < j \leq k < l$ the random vectors $(X_{i+1}, \ldots, X_j)$ and $(X_{k+1}, \ldots, X_l)$ are independent, the *increments* $(Y_l - Y_k)$ and $(Y_j - Y_i)$ of Y are independent.

(d) Y is an $(\mathcal{F}_n)_n$-martingale (special case of Example 6.1).

To obtain a continuous-time process with similar properties, we proceed in two steps: first, fix $N \in \mathbb{N}$, let $\tau = \frac{1}{N}$ and reduce the step length of the random walk to τ by setting $B_0^N = 0$, $B_n^N = \sqrt{\tau} Y_n$ for $n > 0$. This defines a process $B^N : [0, \infty) \to \mathbb{R}$ with domain $\{n\tau : n = 0, 1, \ldots\}$. This *scaled random walk* inherits the above properties of Y, except that $Var B_n^N = \tau n = \frac{n}{N}$. The variance of the *increment* $(B_n^N - B_m^N)$ is thus $\tau(n - m)$.

Fix $s = \frac{m}{N}, t = \frac{n}{N}$. For $p > 0$, define the pth *variation* of the random walk B^N in the interval $[s, t] \subset [0, \infty]$ by

$$V_{[s,t]}^p (B^N) = \sum_{i=m}^{n-1} \left| B_{i+1}^N - B_i^N \right|^p .$$

We compare the *variation* $V_{[s,t]}(B^N)$ and the *quadratic variation* $V_{[s,t]}^2(B^N)$:

Exercise 7.9 Show that $V_{[s,t]}(B^N) \to \infty$ as $N \to \infty$, but that $V_{[s,t]}^2(B^N) = t - s$. (*Hint:* $\left| B_{i+1}^N - B_i^N \right|^p = \tau^{p/2}$!)

When $s = 0$, $V_{[0,t]}^2(B^N) = t$ is usually denoted by $[B^N]_t$ and we can study the *quadratic variation process* $[B^N]$, which reveals itself in the next proposition as the *compensator of the submartingale* $(B_n^N)^2$, in the Doob decomposition described in Chapter 6. Remarkably, the 'increasing process' $[B^N]$ is non-random in this example. (See Exercise 6.12.)

Proposition 7.10 *The process* $X = (X_n)_{n \geq 0}$, $X_n = (B_n^N)^2 - n\tau$ *is a martingale for the discrete filtration* $(\mathcal{F}_{n\tau})_{n \geq 0}$.

Proof $\mathbb{E}[(B_n^N)^2 - (B_m^N)^2 | \mathcal{F}_{m\tau}] = \mathbb{E}[(B_n^N - B_m^N)^2 | \mathcal{F}_{m\tau}]$ for $m < n$, by Lemma 6.11. This is just $\mathbb{E}[(B_n^N - B_m^N)^2] = (n - m)\tau$, since the increment $(B_n^N - B_m^N)$ is independent of $\mathcal{F}_{m\tau}$ and its variance is $(n - m)\tau$, as observed above. Hence

$$\mathbb{E}\left[(B_n^N)^2 - n\tau | \mathcal{F}_{m\tau}\right] = (B_m^N)^2 - m\tau.$$

As we let $N \to \infty$, the domain $\{n\tau : n \le N\}$ increases, so we hope to develop a process defined on $[0, 1]$ which mimics the properties of B^N. To this end, fix $t > 0$ and $\tau = \frac{t}{N}$. For $m\tau = s \le t = N\tau$, we write the value of the scaled random walk B^N at s as

$$B^N(s) = B_m^N = \sqrt{\tau} \sum_{i=1}^{m} X_i.$$

In particular, $B^N(t) = \sqrt{\tau} \sum_{i=1}^{N} X_i$. Consider the sequence $(B^N(t))_{N \in \mathbb{N}}$, keeping t fixed. The well-known *Central Limit Theorem* states that the limit of a sequence of normalised binomial random variables has $\mathcal{N}[0, 1]$ distribution: if $Y_n = \sum_{i=1}^{n} X_i$, where the (X_i) are Bernoulli, then, setting $T_n = \frac{1}{\sqrt{n}} Y_n$, we obtain

$$\lim_{n \to \infty} P(a < T_n < b) = \frac{1}{\sqrt{2\pi}} \int_a^b e^{-\frac{1}{2}x^2} \, dx \text{ for real } a < b.$$

Now $B^N(t) = \sqrt{\tau} Y_N = \sqrt{t} T_N$, so after a simple change of variable we conclude that

$$\lim_{N \to \infty} P(a < B^N(t) < b) = \frac{1}{\sqrt{2\pi t}} \int_a^b e^{-\frac{1}{2t}x^2} \, dx.$$

Applying this argument to the *increments* $B^N(t) - B^N(s)$, with $s = \frac{m}{N}t$, provides a normally distributed limit with mean 0 and variance $(t - s)$. Of course, we only have *convergence in distribution*, and we still need to specify a probability space $(\Omega, \mathcal{F}.P)$ on which the limit can be defined, but we now know what we might demand of the limit process $B = (B_t)_t$ with $B_0 = 0$, as defined in the next section.

7.3 BM: construction and properties

Definition 7.11 Let $(\Omega, \mathcal{F}, P)$ be a probability space. The stochastic process $B : [0, \infty) \times \Omega \to \mathbb{R}$ is a *Brownian Motion* (BM) if:

(a) $B_0 = 0$ a.s.(P);

(b) for $0 \leq s < t < \infty$, $(B_t - B_s) \sim \mathcal{N}(0, t - s)$;

(c) for $m \in \mathbb{N}$ and $0 \leq t_1 < t_2 < \ldots < t_m$, the increments $(B_{t_{i+1}} - B_{t_i})$, $(i < m)$, are independent;

(d) the paths $t \to B_t(\omega)$ are continuous for (P)-almost all $\omega \in \Omega$.

The *standard Brownian filtration* $\mathbb{F} = (\mathcal{F}_t)_{t \geq 0}$ is the augmented filtration (see the discussion preceding Definition 7.4) generated by the process B.

Note that the distribution of the increment $(B_t - B_s)$ depends only on the distance $(t - s)$, so that for any $u > 0$, $(B_{t+u} - B_{s+u})$ has the same distribution as $(B_t - B_s)$; in other words, the increments are *stationary*.

Remark 7.12 We have not yet shown that such a process exists, nor specified on what sort of probability space it can be constructed. There are several possible choices for Ω (see e.g. [K-S],[S]), but we only sketch the construction of the 'canonical' BM on 'path space'. We also focus on the properties of increments, rather than of B itself, for a good reason: we might look instead for a centred process W such that for all $m > 1$ and (strictly increasing) distinct $(t_i)_{i \leq m}$, the random variables $(W_{t_i})_{i \leq m}$ are independent, with joint distribution of $(W_{t_i+t} : i \leq m)$ independent of $t > 0$. But then it turns out that no such process can have a.s. continuous paths! It can be made sense of by means of tempered distributions and is known as 'white noise'; but this is beyond our scope.

By Theorem 3.21, the independence of the increments of B means that for $t_1 < t_2$ the joint density of $B_{t_1} = B_{t_1} - B_0$ and $B_{t_2} - B_{t_1}$ is the product of their individual densities. Writing

$$p(t, x, y) = \frac{1}{\sqrt{2\pi t}} \exp\left(-\frac{(x - y)^2}{2t}\right)$$

for $t > 0$ and real x, y, and with Borel sets A_1, A_2, the joint distribution $P(X_{t_1} \in A_1, X_{t_2} \in A_2)$ equals

$$\int_{A_1} \int_{A_2} p(t_1, 0, x_1) p(t_2 - t_1, x_1, x_2) dx_1 dx_2.$$

The idea now is to treat B as a map $\Omega \to \mathbb{R}^{[0,\infty)}$, i.e. into the vector space of all real functions on $[0, \infty)$, and seek to build a measure ν on this product space by specifying its projections onto *cylinder sets*. These are sets in the product space whose form is determined by finitely many coordinates, i.e. for $n \geq 1$ and a Borel set B in $\mathbb{R}^n$ we consider the set

$$C(B) = \{(\omega(t))_{t \geq 0} \in \mathbb{R}^{[0,\infty)} : (\omega(t_1), \omega(t_2), \dots, \omega(t_n)) \in B\}.$$

This class of sets is generated by taking n-fold products $\prod_{i=1}^{n} [a_i, b_i]$ of closed intervals for the Borel sets B, in which case we have simply specified that $\omega(t_i)$ lies between a_i and b_i for each $i \leq n$. In other words, if we think of B as the describing the continuous motion of a particle on the line, our observations place the particle in the interval $[a_i, b_i]$ at time t_i for $i \leq n$, where $t_1 < \dots < t_n$. The measure we assign to $C\left(\prod_{i=1}^{n} [a_i, b_i]\right)$ is defined via its projection $\nu_{t_1 \dots t_n}$ onto $\mathbb{R}^n$

$$\int_{a_1}^{b_1} \dots \int_{a_n}^{b_n} p(t_1, 0, x_1) \dots p(t_n - t_{n-1}, x_{n-1}, x_n) dx_1 dx_2 \dots dx_n.$$

We call the family of these probability measures (where n ranges over $\mathbb{N}$) the family of *finite-dimensional* (fi-di) *distributions* of B.

Exercise 7.13 Verify that $B = (B_t)_{t \geq 0}$ is a BM iff $B_0 = 0$, almost all paths of B are continuous and the fi-di distributions are given as above.

To extend ν from cylinders to the σ-field $\mathcal{B}^{[0,\infty)}$ generated by the cylinders will require that the fi-di distributions should be *consistent*: for $n \in \mathbb{N}$, real Borel sets A_i, $i \leq n$ and $t_1 < \dots < t_n$ in $[0, \infty)$ we need:
(i) $\nu_{t_1 \dots t_n}(A_1 \times \dots \times A_n) = \nu_{t_1 \dots t_n, t_{n+1}}(A_1 \times \dots \times A_n \times \mathbb{R})$ for any $t_{n+1} \neq t_i$ ($i \leq n$).
(ii) $\nu_{t_{\sigma(1)} \dots t_{\sigma(n)}}(A_1 \times \dots \times A_n) = \nu_{t_1 \dots t_n}(A_{\sigma(1)} \times \dots \times A_{\sigma(n)})$ for any permutation σ of the indices $\{1, 2 \dots n\}$.

These conditions are easily seen to be satisfied by our densities. More generally, suppose that for every $n \in \mathbb{N}$ and each finite set of indices $(t_i)_{i \leq n}$ we have a family of probability measures $\mathcal{J}_n = \nu_{t_1 \dots t_n}$ defined on $(\mathbb{R}^n, \mathcal{B}(\mathbb{R}^n))$. If the collection $\mathcal{J} = (\mathcal{J}_n)_{n \in \mathbb{N}}$ satisfies (i) and (ii), we call it a *consistent family of fi-di distributions*.

Our measure and process are then given by *Kolmogorov's Consistency Theorem* – its proof, based on the Caratheodory Extension Theorem, is omitted. (See e.g. [KS].)

Theorem 7.14 *Let $\mathcal{J}$ be a consistent family of fi-di distributions. Then there is a probability measure ν on* $(\mathbb{R}^{[0,\infty)}, \mathcal{B}^{[0,\infty)})$ *and a coordinate process X on this space with $X_t(\omega) = \omega(t)$ such that*

$$\nu(X_{t_1} \in A_1, \ldots, X_{t_n} \in A_n) = \nu_{t_1 \ldots t_n}(A_1 \times \ldots \times A_n).$$

This provides a candidate for BM, but we have no information about its paths.

First, we derive some familiar properties of the BM defined in Definition 7.11. Recall that the *characteristic function* of a random variable X is given by $\phi(\lambda) = \mathbb{E}[e^{i\lambda X}]$ for real λ (i is the imaginary unit). ϕ determines the moments of X via the formula $i^n \mathbb{E}[X^n] = \frac{d^n}{d\lambda^n}|_{\lambda=0}(\phi(\lambda))$ – i.e. where the derivatives are computed at $\lambda = 0$.

Proposition 7.15 *BM $B = (B_t)_{t \geq 0}$ has the following properties for $s, t \geq 0$:*

(i) $B_t \sim \mathcal{N}(0, t)$;

(ii) $Cov(B_s, B_t) = \min(s, t)$;

(iii) B_t has characteristic function $\phi(\lambda) = e^{-\frac{1}{2}\lambda^2 t}$, $\lambda \in \mathbb{R}$;

(iv) $\mathbb{E}\left[B_t^4\right] = 3t^2$.

Proof (i) is immediate, as the density of $B_t = B_t - B_0$ is $f_{B_t}(x) = \frac{1}{\sqrt{2\pi t}} \exp\left(-\frac{x^2}{2t}\right)$ For (ii), we have, with $s \leq t$

$$Cov(B_s, B_t) = \mathbb{E}[(B_t - B_s + B_s)B_s] = 0 + \mathbb{E}\left[B_s^2\right] = s$$

since $(B_t - B_s)$ and B_s are independent, hence uncorrelated (Exercise 5.14). For (iii) complete the square in the exponent

$$\int_{-\infty}^{\infty} e^{i\lambda x} e^{-\frac{x^2}{2t}} dx = e^{-\frac{1}{2}\lambda^2 t} \int_{-\infty}^{\infty} e^{-\frac{1}{2}\left(\frac{x - i\lambda t}{\sqrt{t}}\right)^2} dx = \sqrt{2\pi t}\, e^{-\frac{1}{2}\lambda^2 t}$$

since $\int_{-\infty}^{\infty} e^{-\frac{1}{2}y^2} dy = \sqrt{2\pi}$, so that $\mathbb{E}[e^{i\lambda X}] = e^{-\frac{1}{2}\lambda^2 t}$. Now (iv) follows on applying the above moment formula.

Recall: a *random vector* is a measurable map $\mathbf{X} = (X_i)_{i \leq d}$ from Ω to $\mathbb{R}^d$, with mean vector $\mu = (\mathbb{E}[X_i])_{i=1,\ldots,d}$ and covariance matrix

$\mathbf{V} = [\sigma_{ij}]_{i,j \leq d}; \sigma_{ij} = Cov(X_i, X_j)$. We call $\mathbf{X}$ *multivariate Gaussian* $\mathcal{N}(\mu, \mathbf{V})$ if its joint density is given by

$$\frac{1}{(2\pi)^{d/2}} (\det \mathbf{V})^{-\frac{1}{2}} \exp\left[-\frac{1}{2}(\mathbf{x} - \mu)^T \mathbf{V}^{-1}(\mathbf{x} - \mu)\right]$$

for $\mathbf{x} \in \mathbb{R}^d$. (Here T denotes the transpose.) If $\mathbf{V}$ is diagonal (i.e. $Cov(X_i, X_j) = 0$ for $i \neq j$), then the form of the density shows that the (X_i) are independent (and the converse holds trivially). Now (ii) above leads at once to a further characterisation of BM. We call a stochastic process $X = (X_t)_{t \geq 0}$ a *Gaussian process* if all its fi-di distributions are multivariate Gaussian. It is *centred* if $\mathbb{E}[X_t] = 0$ for all t.

Exercise 7.16 Show that a centred Gaussian process X on $[0, T]$ with $X_0 = 0$ a.s. is a BM on $[0, T]$ iff (i) it has a continuous version and (ii) $Cov(X_s, X_t) = \min(s, t)$ for all $0 \leq s, t \leq T$. (*Hint:* use (ii) to show that the increments are uncorrelated.)

Use this characterisation to verify the following *symmetry properties of BM*; if B is BM, so are:

(i) $-B$ (reflection),

(ii) $\left(\frac{1}{c} B_{c^2 t}\right)_t$ for $c > 0$ (scaling),

(iii) $(B_{t+u} - B_u)_t$ for $u > 0$ (stationarity),

(iv) $(Z_t = t B_{1/t}$ if $t > 0, Z_0 = 0)$ (inversion) (Warning: Proving continuity at 0 is rather more challenging than the rest, and may be safely omitted here! The brave student should consult [S, p. 59] for a nice approach.)

Return to consider the paths of B: a first thought might be that the set $C[0, \infty)$ of continuous functions $[0, \infty) \to \mathbb{R}$ might be a full set for P, but this is hopeless, as $C[0, \infty)$ is not even in the σ-field $\mathcal{B}^{[0,\infty)}$. (All sets in this σ-field are determined by countably many coordinates.) However, we can use Proposition 7.15(iv) to show that B has a version with a.s. continuous paths; for this, *Kolmogorov's Continuity Theorem* – whose proof is also omitted, see [K-S] – does the trick:

Theorem 7.17 *Let* $X = (X_t)_{t \geq 0}$ *be a stochastic process and* $T > 0$. *If there exist positive* α, β, C *such that for all* s, t *in* $[0, T]$, $\mathbb{E}[|X_t - X_s|^\beta] \leq C |t - s|^{1+\alpha}$, *then* X *has a version with a.s continuous paths.*

This version can be taken as locally Hölder-continuous in that for $\gamma \in \left(0, \frac{\alpha}{\beta}\right)$ *we have* $|X_t - X_s| \le C |t - s|^\gamma$.

For BM B, we take $C = 3, \beta = 4$ and $\alpha = 1$ to conclude that B has a continuous version B^T on $[0, T]$ for any $T > 0$. Now define $\Omega_T = \{\omega : B_r^T(\omega) = B_r(\omega), r \in [0, T] \cap \mathbb{Q}\}$, then Ω_T is in $\mathcal{B}^{[0,\infty)}$, since $B_r(\omega) = \omega_r$, so that Ω_T is determined by countably many coordinates. Also, $\nu(\Omega_T) = 1$ since B^T is a version of B. On the set $\Omega = \cap_{T \in \mathbb{N}} \Omega_T$, which has $\nu(\Omega) = 1$, the random variables B_r^S, B_r^T coincide at each rational $r \le \min(S, T)$ and both are continuous, hence they coincide on $[0, \min(S, T)]$ for every $\omega \in \Omega$. For each fixed $t \ge 0$, denote this common value by $\overline{B}_t(\omega)$ if $\omega \in \Omega$ and set $\overline{B}_t(\omega) = 0$ if $\omega \notin \Omega$. The process $(\overline{B}_t)_{t \ge 0}$ satisfies the definition of BM. For simplicity, we write $B = (B_t)_{t \ge 0}$ for $(\overline{B}_t)_{t \ge 0}$ and P instead of ν, and refer simply to a BM B defined on a probability space $(\Omega, \mathcal{F}, P)$ without specifying the construction. With the standard Brownian filtration $\mathbb{F} = (\mathcal{F}_t)_{t \ge 0}$ (see Definition 7.11), we obtain the filtered space $(\Omega, \mathcal{F}, \mathbb{F}, P)$.

7.4 Martingale properties of BM

BM gives rise to some fundamental $\mathbb{F}$-martingales:

Proposition 7.18 *Let* $B = (B_t)_{t \in [0,T]}$ *be a BM on the filtered space* $(\Omega, \mathcal{F}, \mathbb{F}, P)$. *The following are* $(\mathbb{F}, P)$-*martingales:*
(i) B;
(ii) $X = (X_t)_{t \ge 0}$, *where* $X_t = B_t^2 - t$ *(hence* $t \to t$ *is the compensator of the submartingale* B^2*);*
(iii) $Y = (Y_t)_{t \ge 0}$, *where* $Y_t = \exp\left(cB_t - \frac{1}{2}c^2 t\right)$, *where* $c \in \mathbb{R}$.

Proof For $0 \le r \le s \le t$, the increments $(B_t - B_s)$ and $B_r - B_0 = B_r$ are independent. But $\mathcal{G}_s$, hence $\mathcal{F}_s$, is generated by $\{B_r : r \le s\}$ so $(B_t - B_s)$ is independent of any $\mathcal{F}_s$-measurable function, hence of $\mathcal{F}_s$. So $\mathbb{E}[(B_t - B_s)|\mathcal{F}_s] = 0$. For (ii), we know from Remark 7.6 that for $s \le t$

$$\mathbb{E}\left[\left(B_t^2 - B_s^2\right)|\mathcal{F}_s\right] = \mathbb{E}[(B_t - B_s)^2|\mathcal{F}_s] = \mathbb{E}[(B_t - B_s)^2] = t - s.$$

Finally, (iii) follows from Definition 7.11(c)

$$\mathbb{E}[Y_t|\mathcal{F}_s] = \exp\left(cB_s - \frac{1}{2}c^2t\right)\mathbb{E}[\exp(c(B_t - B_s))|\mathcal{F}_s]$$

$$= \exp\left(cB_s - \frac{1}{2}c^2t\right)\mathbb{E}[\exp(c(B_t - B_s))]$$

$$= \exp\left(cB_s - \frac{1}{2}c^2t\right)\exp\left(\frac{1}{2}c^2(t - s)\right) = Y_s$$

since $c(B_t - B_s) \sim \mathcal{N}(0, c^2(t - s))$.

Remark 7.19 Paul Lévy proved a remarkable 'converse' to Proposition 7.18, which characterises BM among path-continuous martingales:

Lévy's Theorem: Given a path-continuous martingale $M = (M_t)_t$ with $M_0 = 0$ and such that $N = (N_t)_t$, where $N_t = M_t^2 - t$, is again a martingale, then M is a Brownian Motion.

We do not have all the machinery to prove this, but the following computation illustrates the idea behind the proof.

Recall from the discussion preceding Proposition 7.15 that the characteristic function $\phi(\lambda) = \mathbb{E}[e^{i\lambda X}]$ determines the distribution of the random variable X, so that to prove that for $s < t$ the increment $M_t - M_s \sim \mathcal{N}(0, t - s)$, it will suffice to show that

$$\mathbb{E}(e^{iu(M_t - M_s)}) = e^{-\frac{1}{2}u^2(t-s)}.$$

Being normally distributed, the increments of M are independent iff they are uncorrelated, and the latter follows at once from the martingale property of M. Thus Lévy's Theorem will follow if we show that the characteristic function of M has the above form.

Using Taylor's formula, we can write

$$e^{iu(M_t - M_s)} = 1 + iu(M_t - M_s) - \frac{1}{2}u^2(M_t - M_s)^2 + \varepsilon.$$

Take expectations, and recall that $\mathbb{E}[M_t - M_s] = 0$

$$\mathbb{E}\left[1 + iu(M_t - M_s) - \frac{1}{2}u^2(M_t - M_s)^2 + \varepsilon\right]$$

$$= 1 - \frac{1}{2}u^2\mathbb{E}[M_t - M_s]^2 + \varepsilon'.$$

Now $\mathbb{E}[(M_t - M_s)^2] = \mathbb{E}M_t^2 - 2\mathbb{E}[M_tM_s] + \mathbb{E}M_s^2$, and $\mathbb{E}\left(M_t^2\right) = t$ since $N_t = M_t^2 - t$ is a martingale null at 0, while

$$\mathbb{E}[M_tM_s] = \mathbb{E}[\mathbb{E}(M_tM_s|\mathcal{F}_s)] = \mathbb{E}[M_s\mathbb{E}(M_t|\mathcal{F}_s)]$$
$$= \mathbb{E}\left[M_s^2\right] = s,$$

which shows that $\mathbb{E}[M_t - M_s]^2 = t - s$.

For a complete proof, we would have to show that the remainder ε' can be recast suitably to yield

$$1 - \frac{1}{2}u^2(t-s) + \varepsilon' = e^{-\frac{1}{2}u^2(t-s)},$$

which may seem plausible, but is rather technical. We discuss an alternative approach to Lévy's result at the end of Chapter 8.

The exponential martingale $Y_t = \exp\left(cB_t - \frac{1}{2}c^2t\right)$ of Proposition 7.18(iii) plays a fundamental role in applications, as we will confirm in Chapter 8. We shall use it (following [S]) to arrive at a change of the underlying probability measure which turns a 'Brownian Motion with drift', i.e. $X_t = B_t + ct$ for $t \in [0, T]$, into a genuine BM.

First compute expectations of the form $\mathbb{E}[f(X_{t_i}, X_{t_2}, \dots, X_{t_n})]$, where $f : \mathbb{R}^n \to \mathbb{R}$ is a bounded Borel function, and $0 = t_0 < t_1 < \dots < t_n = T$. Since we can always find g such that

$$f(x_1, x_2, \dots, x_n) = g(x_1, x_2 - x_1, \dots, x_n - x_{n-1}),$$

it suffices to consider the density of the random vector

$$\mathbf{X} = (X_{t_1}, X_{t_2} - X_{t_1}, \dots, X_{t_n} - X_{t_{n-1}}),$$

which will allow us to exploit the independence of Brownian increments. To ease the notation, write $\Delta_i = t_i - t_{i-1}$, $\Delta_i x = x_{t_i} - x_{t_{i-1}}$ (with $x_0 = 0$), and similarly for $\Delta_i X$ and $\Delta_i B$. The density of $\mathbf{X}$ has the form $k \prod_{i=1}^{n} \exp \frac{1}{2\Delta_i}(-\Delta_i x - c\Delta_i)^2$, with normalising constant $k = \dfrac{1}{\sqrt{(2\pi)^n t_1 \Delta_1 \Delta_2 \dots \Delta_n}}$. Expanding the square, the density can be written as

$$k \prod_{i=1}^{n} \exp\left(-\frac{(\Delta_i x)^2}{2\Delta_i}\right) \prod_{i=1}^{n} \exp\left(c\Delta_i x - \frac{1}{2}c^2\Delta_i\right)$$
$$= k \prod_{i=1}^{n} \exp\left(-\frac{(\Delta_i x)^2}{2\Delta_i}\right) \exp\left(cx_{t_n} - \frac{1}{2}c^2t_n\right)$$

since $\displaystyle\prod_{i=1}^{n} \exp\left(c\Delta_i x - \tfrac{1}{2}c^2\Delta_i\right) = \exp\left[c\sum_{i=1}^{n} \Delta_i x - \tfrac{1}{2}c^2\sum_{i=1}^{n}\Delta_i\right]$.

But $k\displaystyle\prod_{i=1}^{n} \exp\left(-\frac{(\Delta_i x)^2}{2\Delta_i}\right)$ expresses the joint density of the Brownian increments, so that (independently of the choice of the $(t_i)_{i\leq n}$) we obtain

$$\mathbb{E}[f(X_{t_1}, X_{t_2}, \ldots, X_{t_n})] = \mathbb{E}[f(B_{t_1}, B_{t_2}, \ldots, B_{t_n})]Y_T,$$

where $Y_t = \exp\left(cB_t - \tfrac{1}{2}c^2 t\right)$ is the martingale defined in Proposition 7.18 (iii). This is known as the *tilting formula*, since it 'tilts' the probabilities of the Brownian paths towards those of a BM with 'drift' c. (To go the other way, replace c by $-c$.)

To use this formula to effect a useful change of measure, we must return to the abstract point of view described in Section 7.1, under which a (path-)continuous stochastic process is regarded as a point in the space $C[0, T]$ (with its sup norm). This led to the canonical construction of BM as a process defined on $C[0, T]$, equipped with its Borel σ-field $\mathcal{B}^{[0,T]}$, and $B_t(\omega) = \omega(t)$ for $\omega \in C[0, T]$ with $\omega(0) = 0$. Any continuous stochastic process $X = (X_t)_{t\leq T}$, defined on some (unspecified) probability space $(\Omega', \mathcal{F}', \mu')$ induces a probability measure on $(C[0, T], \mathcal{B})$ by setting $\mu(A) = \mu'(X^{-1}(A))$. We say that μ *corresponds* to the process X.

In particular, let P be the probability corresponding to the standard BM B on $(C[0, T], \mathcal{B})$ and suppose that Q corresponds to the process X above, with $X_t = B_t + ct$. We claim that, for any bounded Borel function Z defined on $C[0, T]$ we have (with expectations under P, Q denoted via subscripts)

$$\mathbb{E}_Q[Z] = \mathbb{E}_P[ZY_T].$$

The proof employs a standard monotone class argument: first, since bounded Borel functions are a.e. limits of simple functions and integration is linear, we only need to check the identity for $Z = \mathbf{1}_A$, where $A \in \mathcal{B}$. First let A have the form $A = \{\omega : a_i \leq \omega(t_i) \leq b_i, i \leq n\}$ for a partition $(t_i)_{i\leq n}$ as above and real $a_i < b_i, i \leq n$. The tilting formula shows that the desired identity holds for $\mathbf{1}_A$ if A is such a set. The class $\mathcal{S}$ of these sets is a π-system and if $\mathcal{C}$ is the collection of sets A for which $Z = \mathbf{1}_A$ satisfies the identity, then $\mathcal{C}$ is a d-system containing $\mathcal{S}$. Thus the identity holds for all A in $\sigma(\mathcal{S})$, and therefore for all Borel sets.

What all this amounts to is summarised in the following result. Note that our arguments require the processes to be defined on $C[0, T]$. In Chapter 8 we outline a proof, based on Lévy's Theorem and martingale calculus, which avoids this restriction. We state the result in the general form.

Lemma 7.20 *(Girsanov's Theorem – special case) Suppose that B is Brownian Motion on the filtered space $(\Omega, \mathcal{F}, \mathbb{F}, P)$ and that the probability measure Q is defined by $\frac{dQ}{dP} = \exp\left(-cB_T - \frac{c^2}{2}T\right)$ on $\mathcal{F} = \mathcal{F}_T$. Then the process B^* defined by $B_t^* = ct + B_t$ is a Brownian Motion on $(\Omega, \mathcal{F}, \mathbb{F}, Q)$.*

7.5 Variation of BM

In Chapter 6 we showed how linear combinations of discrete-time predictable processes and martingale differences provide new martingales which we described as analogous to 'stochastic integrals'. To explore this idea for BM, we must first examine how 'wild' typical paths of BM can be. First, observe that $\mathbb{E}\left[\left(\frac{B_{t+h}-B_t}{h}\right)^2\right] = \frac{1}{h}$ for $h > 0$. Hence if $h \to 0$, the 'differential ratios' diverge in L^2-norm at each t. In fact (although our argument does not prove it!), it can also be shown that almost all paths are nowhere differentiable. (See [K-S] for the rather technical proof.) So we begin by examining the *variability* of the paths, guided by the special case of the scaled random walk B^N discussed in Section 7.2.

Definition 7.21 Fix $T > 0$, a process $X = (X_t)_{t \in [0,T]}$, $t \leq T, n \in \mathbb{N}$. Let $\pi_n = \{t_i : 0 = t_0 < t_1 < .. < t_n = t\}$ be a partition of $[0, t]$. Write $\Delta_i^n X = (X_{t_{i+1}} - X_{t_i})$, and set

$$V_n^p(t, X) = \sum_{i=0}^{n-1} |\Delta_i^n X|^p$$

for $p > 0$. Now let $V_{[0,t]}^p(X)$, be the limit in probability (if it exists) of $V_n^p(t, X)$ for a sequence of partitions π_n whose *mesh* $\rho^{(n)} = \max |t_{i+1} - t_i|$ goes to 0. $V_{[0,t]}^p(X)$ is the pth *variation* of X on $[0, t]$.

Our interest is in the cases $p = 1$ and $p = 2$. We again denote the *quadratic variation* $V^2_{[0,t]}(X)$ by $[X]_t$. When $X = B^N$, we recover the results of Exercise 7.9. For $X = B$, fix a sequence of partitions π_n with mesh going to 0 and for the subintervals of π_n write $\Delta^n_i = \left(t^n_{i+1} - t^n_i\right)$. Now $\Delta^n_i B \sim \mathcal{N}\left(0, \Delta^n_i\right)$, so that $\mathbb{E}[(\Delta^n_i B)^2] = \Delta^n_i$, and by Proposition 7.15(iv), $\mathbb{E}[(\Delta^n_i B)^4] = 3\left(\Delta^n_i\right)^2$. Using the independent centred random variables $X^n_i = (\Delta^n_i B)^2 - \Delta^n_i$, we have

$$\mathbb{E}[(X^n_i)^2] = Var\left((\Delta^n_i B)^2\right) = \mathbb{E}[(\Delta^n_i B)^4] - (\Delta^n_i)^2 = 2\left(\Delta^n_i\right)^2.$$

This leads to:

Proposition 7.22 *Brownian Motion B has infinite variation, but finite quadratic variation, on $[0, T]$. In fact, $[B]_t = t$ ($t \in [0, T]$).*

Proof Write

$$\mathbb{E}\left[\left(V^2_n(t, B) - t\right)^2\right] = \mathbb{E}\left[\sum_{i=0}^{n-1}\left((\Delta^n_i B)^2 - \Delta^n_i\right)^2\right].$$

By definition of X^n_i, the RHS equals $\mathbb{E}\left[\left(\sum_{i=0}^{n-1} X^n_i\right)^2\right] =$

$$\sum_{i=0}^{n-1}\mathbb{E}\left[(X^n_i)^2\right] + \sum_{i \neq j}\mathbb{E}(X^n_i)\mathbb{E}(X^n_j) = 2\sum_{i=0}^{n-1}(\Delta^n_i)^2 \leq 2\rho^{(n)}t,$$

and this goes to 0 with the mesh size of π_n. Convergence in L^2-norm implies convergence in probability, so the quadratic variation of B on $[0, t]$ is $[B]_t = t$.

Note that $V^2_n(t, B) \leq (\max_{i<n}\left|B_{t_{i+1}} - B_{t_i}\right|).V_n(t, B)$, and by continuity of $t \to B_t$ the first term on the right goes to 0 as $\rho^{(n)} \to 0$. Since the LHS goes to $t > 0$, it follows that $V_n(t, B) \to \infty$.

This result illustrates our problem, but also suggests a possible solution. The problem is that the only 'integrators' $F : [0, T] \to \mathbb{R}$ for which Stieltjes integrals $\int_0^T gdF$ can be defined for all (bounded) continuous real functions on $[0, T]$ are functions of bounded variation, i.e. such that $T_{[0,T]}(F) = \sup V_n(T, F)$ is finite, where the sup is taken over all partitions. This results from an application of the uniform boundedness principle in Functional Analysis, which states that, given a family (T_α) of bounded linear maps from a Banach space E to a normed space and such that for each $x \in E$, $\sup_\alpha ||T_\alpha x||$ is finite, then $\sup_\alpha ||T_\alpha|| < \infty$, i.e. the family is uniformly bounded. (See e.g. [R] for a proof.) In particular, for the Banach space $C[0, T]$ of continuous functions on $[0, T]$, with $||f|| = \sup_{x \in [0,T]} |f(x)|$, and any sequence of partitions $(\pi_n)_n$ of $[0, T]$, the linear functionals $(L_n)_n$ defined by

$$L_n(g) = \sum_{i=0}^{n-1} g(t_i) \, |\Delta_i^n F|$$

for $g \in C[0, T]$ are therefore uniformly bounded. Now the function with $h(t_i) = 1$, when $\Delta_i^n F \geq 0$, $h(t_i) = -1$ otherwise, can be extended by linear interpolation to $h \in C[0, T]$, so that $\sup_n ||L_n|| < \infty$ implies $T_{[0,T]}(F) < \infty$.

One consequence is that an 'integral' $\int_0^1 g_s(\omega) dB(\omega)$ cannot be defined pathwise for all $g \in C[0, 1]$. However, finiteness of the quadratic variation suggests an alternative approach as long as we are content with L^2-norm convergence, and allow only *adapted* integrands. Note that the definition of $h(t)$ above depends on the 'future' value of F at t_{i+1} for any $t \in [t_i, t_{i+1}]$; if we restrict to adapted integrands, we can still hope to define their integral against Brownian Motion.

8

Stochastic integrals

Observation of dynamic phenomena, such as population growth, weather changes, electronic signals, stock market values, etc., invariably suffers from inaccuracy caused by limitations in measurement precision or external disturbances. Mathematically we can treat these as 'random errors' which limit our information of the behaviour of the observed phenomenon. In developing predictive models we seek to introduce the random error term into the dynamic equations of our model. For exponential population growth models, this could take the form $dN_t = N_t(\mu_t dt + \sigma_t dB_t)$, $N_0 = a$, where N_t represents population size at time t, μ_t is the average growth rate and $\sigma_t dB_t$ represents the effect of random fluctuations about the average. In all that follows we assume given a BM B on some filtered space $(\Omega, \mathcal{F}, \mathbb{F}, P)$. We exploit the finiteness of its quadratic variation to make mathematical sense of the dB term. Only an outline of the theory of Itô processes and their calculus can be given here: [B-Z] is a good basic reference, while [S] provides many of the details we omit, as well as describing the context of our main application – the Black–Scholes model in finance – in detail. For the general martingale calculus, touched on in the final section, the compendious [D-M] remains the indispensable bible, well complemented by e.g. [R-W], [R-Y].

8.1 The Itô integral

As for random variables, so for processes:

Definition 8.1 An $\mathbb{F}$-adapted process h on $[0, T] \times \Omega$ is *simple* if for a partition $\pi = \{0 = t_0 < t_1 < \ldots < t_n = T\}$ and $\mathcal{F}_{t_i}$-measurable random variables $(h_i)_{i<n}$, h_t satisfies

$$h_t(\omega) = \sum_{i=0}^{n-1} h_i(\omega) \mathbf{1}_{(t_i, t_{i+1}]}(t) \text{ for } 0 \leq t \leq T, \ \omega \in \Omega.$$

Definition 8.2 The *Itô integral* of the simple process h is defined as

$$\int_0^T h_s dB_s = \sum_{i=0}^{n-1} h_i \Delta_i^\pi B = \sum_{i=0}^{n-1} h_i (B_{t_{i+1}} - B_{t_i})$$

whenever $h_i \in L^2(\mathcal{F}_{t_i})$ for all $i \leq n$.

We shall follow convention in using 'forward' increments $\Delta_i^\pi B = (B_{t_{i+1}} - B_{t_i})$ of BM, whereas in Chapter 6 we used 'backward' increments, again in line with common usage. As long as we remain consistent in a specific context, this apparent contrast causes no difficulty: note that in Chapter 6 the 'integrand' was, as here, measurable with respect to the 'earlier' σ-field.

The above stochastic integral is simply a transform of the (finite) discrete martingale $(B_{t_i})_{i=0,1,\ldots,n-1}$ and is therefore automatically an $(\mathbb{F}, P)$-martingale for each simple L^2-process h. The vector space $\mathcal{H}^2_{[0,T]}$ of simple L^2-processes is a subset the following vector space of measurable processes $f = (f_t)_{t \in [0,T]}$

$$\mathcal{M}^2_{[0,T]} = \left\{ f : \mathbb{F}\text{-adapted}, \mathbb{E}\left[\int_0^T f_t^2 dt\right] < \infty \right\}.$$

Note that $\mathcal{M}^2_{[0,T]}$ is a closed subspace of $L^2(dP \times dt)$. One of our main tasks will be to show that $\mathcal{M}^2_{[0,T]}$ is the *closure* of $\mathcal{H}^2_{[0,T]}$ in the L^2-norm. First we note some elementary properties of the integral on $\mathcal{H}^2_{[0,T]}$

Proposition 8.3 *For* $h \in \mathcal{H}^2_{[0,T]}$:

(i) $\mathbb{E}\left[\int_0^T h_t dB_t\right] = 0$;

(ii) (Itô isometry) $\mathbb{E}\left[\left(\int_0^T h_t dB_t\right)^2\right] = \mathbb{E}\left[\int_0^T h_t^2 dt\right]$.

Proof (i) follows as the integral is a martingale transform. The proof of (ii) is identical to the calculation in Section 6.3, but is repeated briefly for emphasis

$$\mathbb{E}\left[\left(\int_0^T h_t dB_t\right)^2\right] = \mathbb{E}\left[\left(\sum_{i=0}^{n-1} h_i \Delta_i^\pi B\right)^2\right]$$

$$= \mathbb{E}\left[\sum_{i,j=0}^{n-1} h_i h_j \Delta_i^\pi B \Delta_j^\pi B\right].$$

As in Section 6.3, the cross terms vanish, since for $i < j$

$$\mathbb{E}\left[h_i h_j \Delta_i^\pi B \Delta_j^\pi B\right] = \mathbb{E}\left[(h_i h_j \Delta_i^\pi B)\,\mathbb{E}\left[\Delta_j^\pi B | \mathcal{F}_{t_j}\right]\right] = 0.$$

So $\mathbb{E}\left[\left(\int_0^T h_t dB_t\right)^2\right] = \sum_{i=0}^{n-1}\mathbb{E}\left[h_i^2 \mathbb{E}\left[(\Delta_i^\pi B)^2 | \mathcal{F}_{t_i}\right]\right]$. But this is $\sum_{i=0}^{n-1}\mathbb{E}\left[h_i^2(t_{i+1}-t_i)\right] = \mathbb{E}\left[\int_0^T h_t^2 dt\right].$

The Itô isometry (ii) is key to the extension of the integral from $\mathcal{H}_{[0,T]}^2$ to $\mathcal{M}_{[0,T]}^2$. On the right we have the square of the $L^2([0,T]\times\Omega)$-norm of the function $h : [0,T]\times\Omega \to \mathbb{R}$, while on the left the integral $I_T(h) = \int_0^T h_t dB_t$ is shown to be an element of $L^2(\Omega)$. As an element of $L^2(dP \times dt)$, $f \in \mathcal{M}_{[0,T]}^2$ has norm $\|f\|_{\mathcal{M}_{[0,T]}^2} = \left(\mathbb{E}\left[\int_0^T f_t^2 dt\right]\right)^{1/2}$, so that (ii) displays the integral operator $I_T : h \to \int_0^T h_t dB_t$ as an *isometry* from $\left(\mathcal{H}_{[0,T]}^2, \|\cdot\|_{\mathcal{M}^2[0,T]}\right)$ into $(L^2(\Omega), \|\cdot\|_2)$, as it implies that $\|I_T(h)\|_2 = \|h\|_{\mathcal{M}_{[0,T]}^2}$. Since the map is continuous, the integral $I_T(f)$ can now be defined uniquely for any f in the *closure* of $\mathcal{H}_{[0,T]}^2$ in the $\mathcal{M}_{[0,T]}^2$-norm, by defining

$$I_T(f) = \lim_{n\to\infty} I_T(h_n), \quad \text{where } h_n \in \mathcal{H}_{[0,T]}^2, \|h_n - f\|_{\mathcal{M}_{[0,T]}^2} \to 0.$$

Exercise 8.4 Verify that the integral on $\mathcal{H}_{[0,T]}^2$ preserves inner products as a map between the Hilbert spaces $\mathcal{M}_{[0,T]}^2$ and $L^2(\Omega)$.

It remains to identify $\mathcal{M}_{[0,T]}^2$ as the closure of $\mathcal{H}_{[0,T]}^2$. We do this in stages:

Lemma 8.5 *Let* $t \to f(t)$ *be continuous as a map* $[0, T] \to L^2(\Omega)$ *and* $\pi_n = (t_i)_{i<n}$ *a refining sequence of partitions of* $[0, T]$ *(recall that this means that* $\pi_n \subset \pi_{n+1}$ *for each* n*). Assume that the mesh* $\rho_n = max_i |t_{i+1} - t_i|$ *of* π_n *goes to* 0 *and* $n \to \infty$*. Set* $h_n(t, \omega) = \sum_{i=0}^{n-1} f(t_i, \omega) \mathbf{1}_{(t_i, t_{i+1}]}(t)$ *on* $[0, T] \times \Omega$*. Then* $h_n \in \mathcal{H}^2_{[0,T]}$ *and* $||f - h_n||_{\mathcal{M}^2_{[0,T]}} \to 0$ *as* $n \to \infty$*.*

Proof Since $t \to f(t)$ is continuous, it is uniformly continuous on $[0, T]$, so $\mathbb{E}[\{f(t) - h_n(t)\}^2]$ can be made less than $\frac{\varepsilon}{T}$ uniformly in t: take n such that $\mathbb{E}[f(t, \omega) - f(t_i, \omega)^2] < \frac{\varepsilon}{T}$ if $t \in (t_i, t_{i+1}], i < n$. Hence $\int_0^T \mathbb{E}[\{f(t) - h_n(t)\}^2] dt < \varepsilon$ for such n.

To prepare for the second step, we show that for *bounded* f in $\mathcal{M}^2_{[0,T]}$, $f_n(t, \omega) = \int_0^\infty e^{-x} f\left(t - \frac{x}{n}, \omega\right) dx$ (where we set $f(s, \cdot) = 0$ if $s \notin [0, T]$) is again in $\mathcal{M}^2_{[0,T]}$. For this, note that the integrand $g(x, t, \omega) = e^{-x} f\left(t - \frac{x}{n}, \omega\right)$ is $\mathcal{B}[0, \infty) \times \mathcal{B}[0, T] \times \mathcal{F}$-measurable, so that by Fubini's Theorem the integral f_n is $\mathcal{B}[0, T] \times \mathcal{F}$-measurable. Similarly, for fixed $t \leq T$ the integrand is $\mathcal{B}[0, \infty) \times \mathcal{F}_t$-measurable, so $\omega \to f_n(t, \omega)$ is $\mathcal{F}_t$-measurable, hence f_n is $\mathbb{F}$-adapted. Finally, since f_n is bounded, $||f_n||_{\mathcal{M}^2_{[0,T]}}$ is finite.

We need a further result about real functions:

Lemma 8.6 *If* $g \in L^2([\mathbb{R}, \mathcal{B}, m)$ *and* $g_s(t) = g(t - s)$ *for real* s, t*, then* $s \to g_s$ *is bounded and uniformly continuous as a map* $\mathbb{R} \to L^2$*.*

Proof The Lebesgue measure m is translation-invariant, therefore $\int_{\mathbb{R}} [g(t - s)]^2 dm(t) = \int_{\mathbb{R}} [g(t)]^2 dm(t)$, so g_s is L^2-bounded. The continuity claim is that given $\varepsilon > 0$ we can find $\delta > 0$ such that $|s - t| < \delta$ implies $||g_s - g_t||_2 < \varepsilon$. To prove this, recall (see e.g. [C-K, Theorem 4.39]) that there is a continuous function h with $h = 0$ off some interval $(-K, K)$ such that $||g - h||_2 < \frac{\varepsilon}{3}$. But h is uniformly continuous on $[-K, K]$, so we can find $0 < \delta < K$ with $|h(s) - h(t)| < \frac{\varepsilon}{3(3K)^{1/2}}$ whenever $|s - t| < \delta$. For such s, t

$$\int_{\mathbb{R}} |h(u - s) - h(u - t)|^2 \, du \leq \left(\frac{\varepsilon}{3}\right)^2 \frac{2K + \delta}{3K},$$

hence

$$||g_s - g_t||_2 \leq ||g_s - h_s||_2 + ||h_s - h_t||_2 + ||h_t - g_t||_2$$
$$\leq ||h_s - h_t||_2 + 2\,||g - h||_2 < \varepsilon,$$

where the final inequality again results from translation-invariance of m.

Now we can prove the key lemma.

Lemma 8.7 *The* $||\cdot||_{\mathcal{M}^2_{[0,T]}}$ *closure of* $\mathcal{H}^2_{[0,T]}$ *includes all bounded functions in* $\mathcal{M}^2_{[0,T]}$.

Proof If f is bounded and in $\mathcal{M}^2_{[0,T]}$ and f_n is as above, then by Jensen's inequality and with $y = x/n$

$$\mathbb{E}\left[\left|\int_0^\infty e^{-x}\left\{\left(f\left(t+s-\frac{x}{n}\right) - f\left(t-\frac{x}{n}\right)\right)\right\}dx\right|^2\right]$$
$$\leq \mathbb{E}\left[\int_0^\infty e^{-x}\left|f\left(t+s-\frac{x}{n}\right) - f\left(t-\frac{x}{n}\right)\right|^2 dx\right]$$
$$\leq n\mathbb{E}\left[\int_0^\infty |f(t+s-y)| - f(t-y)|^2 dy\right]$$
$$\to 0 \text{ as } s \to 0 \text{ (by uniform continuity).}$$

So f_n is continuous and thus in the closure of $\mathcal{H}^2_{[0,T]}$. To prove that f is also in the closure, we need to show that

$$||f - f_n||^2_{\mathcal{M}^2_{[0,T]}} = \mathbb{E}\left[\int_0^T \left|f(t) - \int_0^\infty e^{-x} f\left(t - \frac{x}{n}\right)dx\right|^2 dt\right] \to 0$$

as $n \to \infty$. But since $\int_0^\infty e^{-x}dx = 1$, we need only estimate, using Jensen, and writing $F(t,z) = [f(t) - f(t-z)]$

$$\mathbb{E}\left[\int_0^T \left|\int_0^\infty e^{-x} F\left(t, \frac{x}{n}\right)dx\right|^2 dt\right]$$
$$\leq \mathbb{E}\left[\int_0^T \int_0^\infty e^{-x}\left|F\left(t, \frac{x}{n}\right)\right|^2 dx dt\right]$$
$$= \mathbb{E}\left[\int_0^\infty e^{-x}\left\{\int_0^T \left|F\left(t, \frac{x}{n}\right)\right|^2 dt\right\}dx\right]$$

and the inner integral in the last expression goes to 0 as $n \to \infty$, by Lemma 8.6, and by the DCT so does the expression itself. Thus we have:

Theorem 8.8 $\mathcal{M}^2_{[0,T]}$ *is the* $||\cdot||_{\mathcal{M}^2_{[0,T]}}$-*norm closure of* $\mathcal{H}^2_{[0,T]}$.

Proof Let $f \in \mathcal{M}^2_{[0,T]}$ be given and truncate it at n, so that $f_n = f$ when $|f| \leq n$ and 0 otherwise. We know from Exercise 3.7 that this does not destroy any measurability. As f_n is bounded it is in the closure of $\mathcal{H}^2_{[0,T]}$. Also, by Fubini and the DCT

$$||f_n - f||_{\mathcal{M}^2_{[0,T]}} = \int_0^T \mathbb{E}[(f_n(t) - f(t))^2]$$

$$= \int_0^T \int_{\{|f(t)|>n\}} (f_n(t) - f(t))^2 dP dt \to 0 \text{ as } n \to \infty.$$

We summarise all this in the following definition:

Definition 8.9 The *Itô integral* $I_T(f) = \int_0^T f_t dB_t$ is defined for f in $\mathcal{M}^2_{[0,T]}$ as the unique extension of $I_T(h_n)$, via the Itô isometry: if $(h_n)_n$ in $\mathcal{H}^2_{[0,T]}$ satisfies

$$\lim_{n\to\infty} ||h_n - f||_{\mathcal{M}^2_{[0,T]}} = 0,$$

then $I_T(f) := \lim_n I_T(h_n)$.

The uniqueness of the limit is clear: for, if (h_n) and (g_n) are sequences in $\mathcal{H}^2_{[0,T]}$ converging to $f \in \mathcal{M}^2_{[0,T]}$, then so does the sequence $g_1, h_1, g_2, h_2 \ldots$. Hence the sequence

$$I_T(g_1), I_T(h_1), I_T(g_2), I_T(h_2), \ldots$$

of its integrals has an L^2-limit, and thus all its subsequences have the same limit; in particular, $\lim_n(I_T(g_n)) = \lim_n I_T(hn)$.

The main properties of the integral extend easily:

Theorem 8.10 *For f in* $\mathcal{M}^2_{[0,T]}$:

(i) $\mathbb{E}\left[\int_0^T f_t dB_t\right] = 0$;

(ii) $\mathbb{E}\left[\left(\int_0^T f_t dB_t\right)^2\right] = \mathbb{E}\left[\int_0^T f_t^2 dt\right]$.

Proof For simple functions, $(h_n)_n$ (i) holds by Proposition 8.3. More-over, $\mathbb{E}\left[\left(\int_0^T h_n(t)dB_t\right) - \int_0^T f_t dB_t\right]\right)^2$ converges to 0 for some sequence of simple functions $(h(n))_n$. But L^2-convergence implies L^1-convergence, so $\mathbb{E}\left[\int_0^T f_t dB_t\right] = 0$.

(ii) Isometry is preserved in the limit: $||I_T(h_n)||_2 = ||h_n||_{\mathcal{M}^2_{[0,T]}}$ for all n, so $||I_T(f)||_2 = ||f||_{\mathcal{M}^2_{[0,T]}}$ if $h_n \to f$.

8.2 The integral as a martingale

We consider the integral over $[0, t]$ for any $t \leq T$ and seek to con-struct a *stochastic process* $(t, \omega) \to I_t(f)(\omega) = \left(\int_0^t f_u dB_u\right)(\omega)$ on $[0, T] \times \Omega$, for any fixed $f \in \mathcal{M}^2_{[0,T]}$. But now we have to contend with the fact that $I_t(f)$, as in Definition 8.9, is defined a.s.(P) for each t, and the union of the exceptional sets need not be 0. So we do not yet have a definition of a viable process $(I_t(f))_{t \in [0,T]}$. Path-continuity properties will help to overcome this, as we can then restrict attention to a countable dense set of indices in $[0, T]$ (e.g. rationals). Note that $\mathbf{1}_{[0,t]}f : (s, \omega) \to \mathbf{1}_{[0,t]}(s)f_s(\omega)$ is in $\mathcal{M}^2_{[0,T]}$, since the measurabil-ity properties of f remain intact, and $||\mathbf{1}_{[0,t]}f||_{\mathcal{M}^2_{[0,T]}} \leq ||f||_{\mathcal{M}^2_{[0,T]}}$. Thus $I_T(\mathbf{1}_{[0,t]}f) = \int_0^T \mathbf{1}_{[0,t]}(s)f dB_s$ is well-defined. We now show that $I_T(\mathbf{1}_{[0,T}f)$ has a *version* $M = (M_t)_{t \in [0,T]}$ (see Definition 7.1) that is a continuous $\mathbb{F}$-martingale. It is then natural to write $M_t = \int_0^t f_s dB_s$ for $t \in [0, T]$.

Theorem 8.11 *If $f \in \mathcal{M}^2_{[0,T]}$, there is a continuous martingale M such that for each $t \in [0, T]$, $P\left(M_t = \int_0^T \mathbf{1}_{[0,t]}(s)f dB_s\right) = 1$.*

Proof Take $(h^{(n)})_n$ in $\mathcal{H}^2_{[0,T]}$ converging to f in norm, then for each n

$$\mathbb{E}\left[\int_0^T (\mathbf{1}_{[0,t]}(s)h^{(n)}(s))^2 ds\right] = \mathbb{E}\left[\int_0^t [h^{(n)}(s)]^2 ds\right] < \infty,$$

so that $M_t^{(n)} = \int_0^T (\mathbf{1}_{[0,t]}h^{(n)})(s)dB_s$ is well-defined. On the other hand, $\mathbf{1}_{[0,t]}h^{(n)}$ is a simple function. More precisely, if the partition $\pi^n = \{0 = t_0 < t_1 < \ldots < t_{K_n} = T\}$ defines $h^{(n)} = \sum_{i=0}^{K_n} \mathbf{1}_{(t_i, t_{i+1}]}h_i$, with h_i in $L^2(\mathcal{F}_{t_i})$ and $t \in (t_m, t_{m+1}]$, then

$$(\mathbf{1}_{[0,t]}h^{(n)})(s) = \sum_{i=0}^{m-1} \mathbf{1}_{(t_i,t_{i+1}]}(s)h_i(\omega) + \mathbf{1}_{(t_m,t]}(s)h_m(\omega).$$

The Itô integral of $\mathbf{1}_{[0,t]}h^{(n)}$ is, by Definition 8.1

$$M_t^{(n)} = \int_0^T (\mathbf{1}_{[0,t]}h^{(n)})(s)dB_s = \sum_{i=0}^{m-1} h_i(s)\Delta_i^n B + h_m(s)(B_t - B_{t_i})$$

and this is (a.s.(P)) continuous in t since $t \to B_t$ is. To obtain our M_t as a limit of $M_t^{(n)}$, we need *uniform* continuity. We first check that each $\left(M_t^{(n)}\right)_{t \in [0,T]}$ is a martingale – the easy calculation is left as an exercise. Apply the strengthened Doob inequality (see Theorem 7.8) to the positive submartingale $\left|M_t^{(n)} - M_t^{(k)}\right|$ with $p = 2$

$$\lambda^2 P\left(\sup_{t \in [0,T]}\left|M_t^{(n)} - M_t^{(k)}\right| > \lambda\right) \le \mathbb{E}\left[\left(M_T^{(n)} - M_T^{(k)}\right)^2\right].$$

Using the Itô isometry, the RHS becomes

$$\mathbb{E}\left[\left\{\int_0^T (h^{(n)}(s) - h^{(k)}(s))dB_s\right\}^2\right]$$

$$= \mathbb{E}\left[\int_0^T (h^{(n)}(s) - h^{(k)}(s))^2 ds\right],$$

which is just $||h^{(n)} - h^{(k)}||_{\mathcal{M}_{[0,T]}^2}$.

But the sequence $(h^{(n)})_n$ converges to f in L^2 norm, so it is Cauchy in this norm. Hence we can find a subsequence $(n_j)_{j \ge 1}$ with $||h^{(n_j)} - h^{(k)}||_{\mathcal{M}_{[0,T]}^2} < \left(\frac{1}{8}\right)^j$ for all $k > n_j$, $j = 1, 2 \ldots$ So for each $j \ge 1$, $\mathbb{E}\left[\left(M_T^{(n_j)} - M_T^{(k)}\right)^2\right] < \left(\frac{1}{8}\right)^j$ whenever $k > n_j$. Take $\lambda = 2^{-j}$; the Doob inequality gives

$$P\left(\sup_{t \in [0,T]}\left|M_t^{(n_j)} - M_t^{(n_{j+1})}\right| > \frac{1}{2^j}\right) \le \frac{1}{2^j}.$$

Let $A_j = \left\{\omega : \sup_{t \in [0,T]}\left|M_t^{(n_j)}(\omega) - M_t^{(n_{j+1})}(\omega)\right| \le \frac{1}{2^j}\right\}$, so that $\sum_{j \ge 1} P\left(A_j^c\right) \le \sum_{j \ge 1} \frac{1}{2^j} < \infty$, and the first BC Lemma yields $P\left(\limsup_j A_j^c\right) = 0$, and by Exercise 1.10 $P(\liminf_j A_j) = 1$. So for P-almost all ω we find $L(\omega) \in \mathbb{N}$ such that

$\sup_{t \in [0,T]} \left| M_t^{(n_j)}(\omega) - M_t^{(n_{j+1})}(\omega) \right| \leq \frac{1}{2^j}$ for all $j \geq L(\omega)$. So $\left(M_t^{(n_j)}(\omega) \right)_j$ is Cauchy in the norm of $C[0,T]$, hence converges to an element of this Banach space, for a full set of ω. Denote the limit by $M_t(\omega)$ and put $M_t(\omega) = 0$ on the exceptional null set.

Our estimates also show that for each fixed $t \in [0,T]$, the random variables $\left(M_t^{(n_j)} \right)_j$ converge in L^2-norm, and since each $M^{(n)}$ is a martingale, so is M: we have $\mathbb{E}\left[M_t^{(n_j)} | \mathcal{F}_s \right] = M_s^{(n_j)}$ for all j when $s \leq t$, and we need to show that this also holds in the limit. On the right, for each s, the random variables $M_s^{(n_j)}$ converge in L^2-norm, so some subsequence converges a.s., so in the limit we obtain M_s. On the left we have $\mathbb{E}[M_t | \mathcal{F}_s]$ in the limit since $X \to \mathbb{E}[X|\mathcal{G}]$ does not increase the L^2-norm (Proposition 5.23(x)), hence we can again argue as on the right. Finally, we check that the continuous martingale M equals $\int_0^T (\mathbf{1}_{[0,t]} f)(s) dB_s$: $\|\mathbf{1}_{[0,t]} h^{(n_j)} - \mathbf{1}_{[0,t]} f\|_{\mathcal{M}^2_{[0,T]}} \leq \|h^{(n_j)} - f\|_{\mathcal{M}^2_{[0,T]}} \to 0$, so that $\left\| \int_0^T (\mathbf{1}_{[0,t]} h^{(n_j)} - \mathbf{1}_{[0,t]} f(s)) dB_s \right\|_2 \to 0$ by the Itô isometry. But by construction of M_t, $\left\| \int_0^T (\mathbf{1}_{[0,t]} h^{(n_j)}(s) dB_s - M_t \right\|_2 \to 0$. So $M_t = \int_0^T (\mathbf{1}_{[0,t]} f)(s) dB_s$ P-a.s. This completes the proof.

Example 8.12 Since we do not distinguish between different *versions* of a process, this result displays the Itô integral on $[0,T]$ as a linear map I from $\mathcal{M}^2_{[0,T]}$ into $\mathbb{M}^2_c[0,T]$, the vector space of continuous square-integrable martingales. But BM B on $[0,T]$ is a member of $\mathcal{M}^2_{[0,T]}$: (B_t) is a measurable adapted process and, using Fubini

$$\mathbb{E}\left[\int_0^T B_t^2 dt \right] = \int_0^T \mathbb{E}\left[B_t^2 \right] dt = \int_0^T t \, dt = \frac{T^2}{2} < \infty.$$

So it is natural to ask what the random variable $I_t(B) = \int_0^t B_s dB_s$ looks like. The normal rules of calculus would lead to the guess $\frac{1}{2} B_t^2$, but the normal rules do not apply! The correct value is $\int_0^t B_s dB_s = \frac{1}{2} B_t^2 - \frac{1}{2} t$. (You may verify this directly: use the identity $a(b-a) = \frac{1}{2}(b^2 - a^2) - \frac{1}{2}(b-a)^2$ for successive values $a = B_{t_j^{(n)}}, b = B_{t_{j+1}^{(n)}}$ for partitions $\left(t_j^{(n)} \right)_{j \leq n}$ of $[0,T]$, then show that $\sum_{j=0}^{n-1} B_{t_j^{(n)}} \left(B_{t_{j+1}^{(n)}} - B_{t_j^{(n)}} \right)$ converges in L^2-norm to $\frac{1}{2} \left(B_T^2 - T \right)$. With a little effort this leads to our claim. See [B-Z] if you get stuck.) As further evidence, we compare means and variances on the two sides: on the left, $\mathbb{E} \left[\int_0^t B_s dB_s \right] = 0$ as the integral is a centred martingale, and on the right the mean is 0 since $\mathbb{E} \left[B_t^2 \right] = t$. The variances on the two sides also agree:

$$\mathbb{E} \left[\left(\int_0^t B_s dB_s \right)^2 \right] = \mathbb{E} \left[\int_0^t B_s^2 ds \right] = \int_0^t \left(\mathbb{E} \left[B_s^2 \right] \right) ds = \frac{1}{2} t^2 \text{ and}$$

$$\mathbb{E} \left[\left(\frac{1}{4} \left(B_t^2 - t \right)^2 \right) \right] = \frac{1}{4} \left(\mathbb{E} \left[B_t^4 \right] - t^2 \right) = \frac{1}{4} (3t^2 - t^2) = \frac{1}{2} t^2.$$

8.3 Itô processes and the Itô formula

Thus the stochastic integral leads to a different calculus from the normal one, due to the presence of a non-vanishing quadratic variation in the integrator: with $f(x) = x^2$ we obtained $f(B_t) = \int_0^t ds + 2 \int_0^t B_s dB_s$, and similarly with $g(x) = x^3$ we have $g(B_t) = 3 \int_0^t B_s ds + 3 \int_0^t B_s^2 dB_s$ (see e.g. [B-Z]). The processes on the right deserve a name:

Definition 8.13 An *Itô process* $X = (X_t)_{t \in [0,T]}$ is a stochastic process of the form $X_t = X_0 + \int_0^t K_s ds + \int_0^t H_s dB_s$, where X_0 is $\mathcal{F}_0$-measurable, H, K are adapted processes, and $\int_0^T |K_s| \, ds$ and $\int_0^T H_s^2 ds$ are finite a.s.(P).

For convenience, we often write this in differential form: $dX_t = K_t dt + H_t dB_t$, although strictly this is meaningless. If $H \in \mathcal{M}_{[0,T]}^2$, the stochastic integral is well-defined by Definition 8.9.

Remark 8.14 The extension of the integral to all H with a.s. finite integral $\int_0^T H_s(\omega) dB_s(\omega)$ requires some effort, and we briefly sketch the main elements of the *localisation* technique by which this can be done.

The idea is to approximate a path-continuous process X by bounded processes $X^{(n)}$ with $\left| X_t^{(n)}(\omega) \right| \leq n$ for all t and almost all ω. One way of doing this is to wait till X hits an endpoint of $[-n, n]$ and then keep it constant. However, the 'instant' t at which this occurs depends on the chosen path, so is a random variable with values in $[0, \infty]$, called a *stopping time*. Since we work within the finite interval $[0, T]$, we specialise to this case: a random variable $\tau : \Omega \to [0, T]$ is a *stopping time* if $\{\omega : \tau(\omega) \le t\} \in \mathcal{F}_t$ for every $t \le T$. This ensures that the decision on 'when to stop' X is taken on the basis of our knowledge of the paths to X up to time t. We use stopping times to make sense of the stochastic integral for a general Itô process by bringing integrands back to $\mathcal{M}^2_{[0,T]}$: an increasing sequence $(\tau_n)_n$ of stopping times is called $\mathcal{M}^2_{[0,T]}$-*localising for* X if $X\mathbf{1}_{\{t \le \tau_n\}} \in \mathcal{M}^2_{[0,T]}$ and $\tau_n \uparrow T$ a.s.(P), i.e. $P(\cup_n \{\omega : \tau_n(\omega) = T\}) = 1$.

Denote by $\mathcal{L}^2_{loc}[0, T]$ the set of measurable adapted processes H such that $P\left(\int_0^T H_s^2 ds < \infty\right) = 1$. Then $\mathcal{M}^2_{[0,T]} \subset \mathcal{L}^2_{loc}[0, T]$ and for any continuous real function g we have $g(B_t) \in \mathcal{L}^2_{loc}[0, T]$, as B is path-continuous. We can now define the Itô integral when $H_t = g(B_t)$, since the sequence $(\tau_n)_n$, where $\tau_n = \inf\left\{s : \int_0^s H_u^2 du \ge n\right\}$, is $\mathcal{M}^2_{[0,T]}$-localising for H: writing $X_t = \int_0^s H_u^2 du$ we see that $H\mathbf{1}_{\{t \le \tau_n\}} \in \mathcal{M}^2_{[0,T]}$, since $|X_t \mathbf{1}_{\{t \le \tau_n\}}| \le n$ a.s.(P), while $\left\{\int_0^T H_s^2 ds < \infty\right\} = \cup_n\{\tau_n = T\}$.

Fix H in $\mathcal{L}^2_{loc}[0, T]$. For the above localising sequence $(\tau_n)_n$, denote by $M_t^{(n)}$ the continuous martingale version of the Itô integral of $H\mathbf{1}_{\{t \le \tau_n\}}$ and *define* the Itô integral $I_t(H)$ of H as the a.s. limit of the sequence $(M^{(n)})_n$. More precisely, take M as the unique continuous process on $[0, T]$ with $P\left(M_t = \lim_n M_t^{(n)}\right) = 1$ for each t in $[0, T]$. We again write $M_t = I_t(H)$.

We must check that the martingales $M_t^{(n)}$ match up correctly, that the extension of the integral is independent of the localising sequence chosen and that the limit is a continuous process. (See [S] for a nice exposition.) However, the limit process $M = I(H)$ need no longer be a martingale – it is only a *local* martingale in the following sense: a process M adapted to a filtration $\mathbb{F} = (\mathcal{F}_t)_{t \in [0,T]}$ is a *local martingale* if there is an increasing sequence of stopping times (τ_n) with $\tau_n \uparrow T$ a.s. and such that each $M^{(n)}$ defined by $M_t^{(n)} = M_{t \wedge \tau_n}$ is an $(\mathbb{F}, P)$-martingale.

Having described the 'natural home' of the Itô integral, we return to Itô processes and outline the main elements of the resulting Itô calculus. By Theorem 8.11, any Itô process X with H in $\mathcal{M}^2_{[0,T]}$ has a path-continuous version, and if $K = 0$, then $X - X_0$ is a centred martingale.

So if $dX_t = K_t dt + H_t dB_t$ and $\int_0^t \mathbb{E}[|K_s|]ds = \mathbb{E}\left[\int_0^t |K_s|\, ds\right] < \infty$, then $\mathbb{E}[X_t - X_0] = \int_0^t \mathbb{E}[K_s]ds$, since $\mathbb{E}\left[\int_0^t H_s dB_s\right] = 0$.

Recall from Section 6.2 that the Doob decomposition splits a discrete adapted process Y uniquely (up to constants) into a martingale and a predictable process (its *compensator*). With $Y = M^2 = N + A$ for martingales M, N, we found that the compensator A is increasing. The decomposition has a counterpart for continuous-time martingales, where a key issue is the definition of 'predictable' processes (see Section 8.5 and [K]). For our present purposes, it suffices to compute the (continuous-time) compensator of any Itô process M of the form $M_t = \int_0^t H_s dB_s$. First, we note that for $s \leq t$ in $[0, T]$, $\mathbb{E}\left[\left(\int_s^t H_u dB_u\right)^2 |\mathcal{F}_s\right] = \mathbb{E}\left[\left(\int_s^t H_u^2 du\right)|\mathcal{F}_s\right]$. This conditional version of the Itô isometry is similar to Proposition 8.3: for $s < t$ in $[0, T]$, begin with a simple process $H = \sum_{i=0}^{n-1} h_i \Delta_i B$, and assume without loss that $s = t_j, t = t_k$ for some $j, k \leq n$. The LHS is then $\mathbb{E}\left[\left(\sum_{i=j}^{k-1} h_i \Delta_i B\right)^2 |\mathcal{F}_{t_j}\right]$, and equals $\sum_{i=j}^{k-1} \mathbb{E}\left[h_i^2 (\Delta_i B)^2 |\mathcal{F}_{t_j}\right]$ just as before. Since L^2-limits preserve this identity, the result follows for arbitrary H in $\mathcal{M}_{[0,T]}^2$. Now we have:

Theorem 8.15 *If $H \in \mathcal{M}_{[0,T]}^2$ and $M_t = \int_0^t H_s dB_s$, then the process $M^2 - A$, where $A_t = \int_0^t H_s^2 ds$ for $t \in [0, T]$, is also a martingale. We call A the* compensator *of M.*

Proof We obtain

$$\mathbb{E}\left[\left(M_t^2 - M_s^2\right)|\mathcal{F}_s\right] = \mathbb{E}[(M_t - M_s)^2|\mathcal{F}_s] = \mathbb{E}\left[\left(\int_s^t H_u dB_u\right)^2 |\mathcal{F}_s\right]$$

$$= \mathbb{E}\left[\int_s^t H_u^2 du |\mathcal{F}_s\right],$$

from Remark 7.6 and the above conditional form of the Itô isometry. But then $\mathbb{E}\left[\left(M_t^2 - \int_0^t H_u^2 du\right)|\mathcal{F}_s\right] = M_s^2 - \int_0^s H_u^2 du$, as required.

As for B, the compensator of the martingale M can be identified with its *quadratic variation*: i.e. $[M]_t = \int_0^t H_s^2 ds$ for all $t \in [0, T]$ – the proof of this needs some effort (see [S, Th.8.6]) and we omit it. On the other hand, it is easily seen that if $X_t = X_0 + \int_0^t K_s ds$, then $[X]_t = 0$ for all t: P – almost all paths are uniformly continuous on $[0, T]$, so for any sequence (π_n) of partitions of $[0, T]$ with mesh going to 0 we have (using the notation of Definition 7.21) $\sum (\Delta_i^n X)^2 \leq \rho^{(n)} \sum |\Delta_i^n X| \leq \rho^{(n)} \left| \int_0^t K_s ds \right|$ and the RHS goes to 0 with $\rho^{(n)}$.

Corollary 8.16 *The decomposition $dX_t = K_t dt + H_t dB_t$ of an Itô process is unique: if also $dX_t = K_t' dt + H_t' dB_t$, then for all t, $K_t = K_t'$ and $H_t = H_t'$ a.s.(P).*

For then $\int_0^t (K_s - K_s') \, ds = \int_0^t (H_s' - H_s) \, dB_s$ and on the left the quadratic variation is 0, while on the right it is $\int_0^t (H_s' - H_s)^2 \, ds$. Thus $H' = H$ and so $K = K'$ a.s.(P).

The two examples at the beginning of this section provide a brief glimpse of the *Itô formula*, which is the key to the new calculus we have worked towards. In its simplest form, as above with $f(x) = x^2$ or x^3, this states that if $f \in C^2(\mathbb{R})$ (i.e. is twice continuously differentiable), then for $t \in [0, T]$

$$f(B_t) = f(0) + \int_0^t f'(B_s) dB_s + \frac{1}{2} \int_0^t f''(s) ds.$$

This breaks up $f(B_t)$ into the sum of a martingale term and one of bounded variation. The proof of this result is rather technical and will be omitted, but we have already gathered all its essential elements (see [B-Z], Theorem 7.5 for a nicely explicit proof). The same analysis (with more effort) yields an Itô formula for Itô processes, if $dX_t = K_t dt + H_t dB_t$ and $f \in C^{1,2}([0, \infty) \times \mathbb{R})$ (so that $f = f(t, x)$ has continuous partial derivatives f_t, f_x, f_{xx} for $t \in [0, T]$), $f(t, X_t)$ is again an Itô process with unique decomposition

$$f(t, X_t) = f(0, 0) + \int_0^t f_x(s, X_s) H_s dB_s + \int_0^t Y(s, X_s) ds,$$

where $Y(s, X_s) = f_t(s, X_s) + f_x(s, X_s) K_s + \frac{1}{2} f_{xx}(s, X_s) H_s^2$. With this formula we can solve *stochastic differential equations* (SDEs) of the form $dX_t = a(t, X_t) dt + b(t, X_t) dB_t$, $X_0 = x_0$. The solution X we seek is an Itô process with a.s. continuous paths. As for ODEs, an argument using the Contraction Mapping Theorem will provide a unique solution X when f, g are Lipschitz continuous and $X_0 \in L^2(\mathcal{F}_0)$; again, see e.g. [B-Z] for a proof.

We illustrate the impact of the Brownian term by considering equations of the form $dX_t = X_t(\nu dt + \sigma dB_t)$, where the functions a, b above become $a(t, x) = \nu x$, $b(t, x) = \sigma x$, $(\sigma > 0)$, so that the t-derivative f_t vanishes. If instead of B we had a non-random C^1-function $h(t)$ with $h(0) = 0$ and $x_0 > 0$, then the ODE $dX_t = X_t(\nu dt + \sigma dh(t))$ has solution $\log\left(\frac{X_t}{x_0}\right) = \nu t + \sigma h(t)$, so we would have $X_t = x_0 e^{\nu t + \sigma h(t)}$. However, if we apply the Itô formula with $X_t = X_0 e^{\nu t + \sigma B_t}$, i.e. $f(t, x) = x_0 e^{\nu t + \sigma x}$, we obtain

$$d(X_0 e^{\nu t + \sigma B_t}) = \left(\nu + \frac{\sigma^2}{2}\right) X_0 e^{\nu t + \sigma B_t} dt + \sigma X_0 e^{\nu t + \sigma B_t} dB_t.$$

So $X_t = X_0 e^{\nu t + \sigma B_t}$ solves a linear SDE with an adjusted dt-term. We can exploit this, setting $\mu = \nu + \frac{\sigma^2}{2}$, so the above SDE becomes $dX_t = X_t[(\mu dt + \sigma dB_t]$, and has unique solution $X_t = X_0 e^{\nu t + \sigma B_t} = X_0 e^{\left(\mu - \frac{\sigma^2}{2}\right)t + \sigma B_t}$.

8.4 The Black–Scholes model in finance

Our choice of this equation comes from finance. When $\sigma > 0$, $\frac{dS_t}{S_t} = \mu dt + \sigma dB_t$, S_0 constant, describes the stock price dynamics of the famous (even infamous?) *Black–Scholes* (BS) model of mathematical finance; this SDE describes proportional changes in the values (S_t) of an asset (e.g. a stock) in terms of its average *drift* μ and its *volatility* σ, which governs the size of random changes. We see that $S_t = S_0 e^{\left(\mu - \frac{\sigma^2}{2}\right)t + \sigma B_t}$ is then the value of the asset S at any time $t \in [0, T]$. This asset pricing model forms the basis of the theory of *derivative securities*. The risky asset ('stock') S is assumed to have the above BS dynamics, while an accompanying bank account (or riskless bond) β simply satisfies $d\beta_t = r\beta_t dt$, $\beta_0 = 1$, so that $\beta_t = e^{rt}$ accrues at a constant interest rate r.

Our aim is to derive a 'rational price' at time $t < T$ for financial instruments whose value at time T is a function of the final stock price $S_T : Z_T = \Phi(S_T)$. These are *European derivative securities*; the simplest is the *call option*, where $Z_T = (S_T - K)^+$ for a constant K. The call option allows its buyer the right (but not the obligation) to buy a share of the asset S at time T at the *strike price* K, which is determined at time 0. If $S_T \le K$, the option is not exercised, but otherwise the buyer profits by $S_T - K$. Thus the buyer will never lose money at time T, and should therefore have to make some payment to the seller at time 0 to purchase the option, else he would have a chance of a riskless profit (*arbitrage* or 'free lunch'), which an efficient market should not allow. Thus we take as *axiomatic* that *markets do not allow arbitrage*, which can be expressed as follows: for $t \in [0, T]$, let $V_t = a_t S_t + b_t \beta_t$ denote the *value* of a portfolio of bonds and stocks. An *arbitrage opportunity* is a portfolio which starts with 0 a.s.(P), is never negative and has positive probability of being strictly positive at time T, so that $\mathbb{E}[V_T] > 0$. Our axiom excludes such portfolios.

By continuously adjusting our portfolio, we now seek to *replicate* (mimic) the random cashflow process of the call option. It should be done in a *self-financing* manner, so that no cash flows into or out of the total holdings during $(0, T)$. This condition can be expressed in terms of (stochastic) differentials as $dV_t = a_t dS_t + b_t d\beta_t$. The *trading strategy* process $(a_t, b_t)_{t \in [0,T]}$ can make use, at time t, only of knowledge of the stock movements up to that point, i.e. it should be adapted to the Brownian filtration $\mathbb{F}$. Hence the differential equation can be made sense of by Itô integrals, and we add the terminal condition $V_T = Z_T$. In this (idealised) mathematical market model we can determine the initial price C_0 of the option (indeed, its price C_t for any t in $[0, T]$) as well as the strategy (a_t, b_t) uniquely in advance.

The solution of this problem relies on a simple, but beautiful, application of martingale theory. To see this, first express all prices in discounted terms by introducing the discounted prices $\widetilde{S}_t = S_t / \beta_t$, which express the value of our stock in 'time 0' terms. The discounted value of our holdings at time t is, similarly, $\widetilde{V}_t = V_t / \beta_t$, so that $\widetilde{V}_t = a_t \widetilde{S}_t + b_t$.

In the BS model a self-financing trading strategy is a pair of measurable adapted processes (a, b) with value process

$$V_t = V_0 + \int_0^t a_u dB_u + \int_0^t b_u d\beta_u.$$

Exercise 8.17 Use the Itô formula to show that the discounted price satisfies $d\widetilde{V}_t = a_t d\widetilde{S}_t$; in other words, that for $t \in [0, T]$

$$\widetilde{V}_t = V_0 + \int_0^t a_u d\widetilde{S}_u.$$

In the present context we can safely assume that $a = (a_t)_t$ is in $\mathcal{M}^2_{[0,T]}$ and $b = (b_t)_t$ in $L^1(m)$. Thus the exercise shows that, if the discounted price is a martingale, then so is the discounted value process. From $d\widetilde{V}_t = a_t d\widetilde{S}_t$, it follows that changes in discounted value depend only on our stock holdings – the bond plays no role in these. Recall that we are working in some (unspecified) filtered probability space $(\Omega, \mathcal{F}, \mathbb{F}, P)$ such that B is a Brownian Motion under $\mathbb{F}$ – in fact, we shall assume that $\mathbb{F} = (\mathcal{F}_t)_{t \in [0,T]}$ is the augmented Brownian filtration, and that $\mathcal{F}_0$ is trivial (consists just of P-null sets and their complements). The key idea now is to change the probability measure P to a *risk-neutral* measure Q, i.e. a probability measure Q equivalent to P such that the discounted price process $\widetilde{S}$ is an $(\mathbb{F}, Q)$ -martingale. If this can be done, there can be no arbitrage in the market, as we now show.

Suppose that we have a risk-neutral measure $Q \sim P$ and consider a portfolio V with initial value $V_0 = 0$ a.s. (P), hence $a.s.(Q)$. Since $\widetilde{S}$ is a Q-martingale, so is the stochastic integral $\left(V_0 + \int_0^t a_u d\widetilde{S}_u \right)_{t \in [0,T]}$, hence $\mathbb{E}_Q[\widetilde{V}_T | \mathcal{F}_t] = \widetilde{V}_t$ for each $t < T$, and in particular $\mathbb{E}_Q[\widetilde{V}_T] = V_0 = 0$. But the value process is non-negative, so $\widetilde{V}_T = 0$ $a.s.(Q)$, hence also a.s.(P). Thus the market does not allow arbitrage.

Moreover, if $(V_t)_{t \in [0,T]}$ replicates the option, $V_T = (S_T - K)^+$, and the discounted value of the option at time t must be $\widetilde{V}_t$ at all times, so that its *fair* (or *rational*) price at time 0 is

$$V_0 = \mathbb{E}_Q[e^{-rT}(S_T - K)^+ | \mathcal{F}_0] = \mathbb{E}_Q[e^{-rT}(S_T - K)^+].$$

The problem of pricing the option has been reduced to that of finding a risk-neutral measure Q and computing the above expectation. To do this, we shall employ Lemma 7.20, which is a special case of the famous *Girsanov Theorem*.

Since we wish to work with discounted prices, consider the SDE for $\widetilde{S}$: we have $\widetilde{S}_t = \frac{S_t}{e^{rt}} = S_0 e^{\left(\mu - r - \frac{1}{2}\sigma^2\right)t + \sigma B_t}$, satisfying the SDE $\frac{d\widetilde{S}_t}{\widetilde{S}_t} = (\mu - r)t + \sigma B_t$. Using Lemma 7.20 with $c = \frac{\mu - r}{\sigma}$, so that $B_t^* = \left(\frac{\mu - r}{\sigma}\right)t + B_t$, we obtain $d\widetilde{S}_t = \sigma \widetilde{S}_t dB_t^*$. In terms of this new BM, the stock price S is $S_t = S_0 e^{\left(r - \frac{1}{2}\sigma^2\right)t + \sigma B_t^*}$.

Under Q, Lemma 7.20 says that the discounted price process is a stochastic integral of the BM B^*, hence a martingale. The discounted value process is a stochastic integral, hence also a martingale: $\widetilde{V}_t = e^{-rt}V_t = V_0 + \int_0^t a_t \sigma \widetilde{S}_t dB_t^*$.

Thus if a replicating strategy exists in the model, the initial value of the call option should equal the initial investment V_0 in that strategy, so we can compute the rational price of the call option as

$$V_0 = \mathbb{E}_Q[\widetilde{V}_T] = e^{-rT}\mathbb{E}_Q[V_T] = e^{-rT}\mathbb{E}_Q[(S_T - K)^+]$$

$$= \mathbb{E}_Q\left[e^{-rT}\left(S_0 e^{rT - \frac{1}{2}\sigma^2 T + \sigma B_T^*} - K\right)^+\right]$$

and, denoting $\sigma B_T^* - \frac{\sigma^2 T}{2}$ by Y

$$= e^{-rT}\int_{\log\left(\frac{K}{S_0}\right)-rT}^{\infty} (S_0 e^{rT+y} - K)\frac{1}{\sigma\sqrt{2\pi T}} e^{-\frac{(y+\sigma^2 T)^2}{2\sigma^2 T}} dy$$

$$= S_0 \mathcal{N}(d+) - Ke^{-rT}\mathcal{N}(d-).$$

Exercise 8.18 Verify the final step (calculus!) showing that

$$d\pm = \frac{\log\left(\frac{S_0}{K}\right) + \left(r \pm \frac{1}{2}\sigma^2\right)T}{\sigma\sqrt{T}}.$$

The final line yields the Black–Scholes *fair price* for the European call option, which now forms the starting point for an extensive literature on derivative securities (see e.g. [E-K], [S]), which makes essential use of martingales and stochastic integration.

8.5 Martingale calculus

Our brief analysis of the Black–Scholes model has not been entirely 'honest': we have not checked that a replicating strategy actually exists for the call option. To gain understanding of why and how the model provides a unique fair price, it is instructive to approach this question more generally.

The 'Brownian' calculus is the prime example of a much more wide-ranging *martingale calculus*, based on a deep result which extends the Doob decomposition to continuous time. This allows us to define the compensator (and hence quadratic variation) $[M]$ for a path-continuous L^2-martingale (and indeed more widely): this process ensures that $N = M^2 - [M]$ is a martingale. The process $[M]$ now turns out to be of bounded variation and – crucially – it is *predictable*: for continuous-time processes, this concept lies deeper than in the discrete case. As maps on the product space $\mathbb{T} \times \Omega$, any family of stochastic processes generates a σ-sub-field of $\mathcal{B}_{\mathbb{T}} \times \mathcal{F}$, and we can define the *predictable* σ-field Σ_p as that generated by the family of all continuous adapted processes. A predictable process is then one which is measurable with respect to this σ-field.

As in the discrete case, predictable martingales are constant, so that the decomposition $M^2 = N + [M]$ is unique. Thus we can construct stochastic integrals of the form $\int_0^t f_s dM_s$ via simple processes and using the isometry $\mathbb{E}\left[\left(\int_0^t f_s dM_s\right)^2\right] = \mathbb{E}\left[\int_0^t f_s^2 d[M]_s\right]$, where on the right we have a Lebesgue–Stieltjes integral.

The Itô formula extends to this setting. In its simplest form, with $x \to F(x)$ in C^2), it reads

$$dF(M_t) = F'(M_t)dM_t + \frac{1}{2}F''(M_t)d[M]_t.$$

The analogy with Hilbert-space techniques can be taken further: define the *cross-variation* of two martingales M, N as

$$[M, N] = \frac{1}{4}([X + Y] - [X - Y]),$$

then we have the following *integration-by-parts formula*

$$d(M_t N_t) = M_t dN_t + N_t dM_t + [M, N]_t.$$

(The 'extra' term again derives from the fact that the quadratic variations are not 0.)

These ideas were developed much further by the 'Strassbourg School' led by the late Paul-André Méyer and have found varied applications. The wide scope of stochastic integration techniques is illustrated by the use of *semimartingales* as integrators: if $X = X_0 + M + V$, where M is a (local) martingale and V a process (locally) of bounded variation, then we can again define $\int_0^t f_s dX_s$ in much the same way as for Itô processes. (The localisation technique, which applies generally, was outlined for Itô processes in Remark 8.14.)

In fact, semimartingales are the *most general stochastic integrators*: if we demand that $\int_0^t f_n(s)dX_s \to \int_0^t f(s)dX_s$ in probability whenever $f_n(s) \to f(s)$ uniformly on $[0, t]$, then X can be shown to be a semimartingale. In [P], this characterisation is used to define semimartingales and their extensive theory is developed from this perspective.

However, Brownian Motion is never too far away, and plays a central role in the space of continuous martingales. To illustrate this, we outline the proof of Lévy's Theorem. With the above notation we can rephrase the result discussed in Remark 7.19:

Theorem 8.19 (Lévy's Theorem) *Any continuous martingale M with $M_0 = 0$ and $[M]_t = t$ is Brownian Motion.*

As indicated in Remark 7.19, to prove this it suffices to check that the characteristic function of the increments $M_t - M_s$ has the right form, i.e. $\mathbb{E}[e^{i\lambda(M_t - M_s)}] = e^{-\frac{1}{2}\lambda^2(t-s)}$ for $0 \le s < t \le T$ and all real λ. So apply the above Itô formula to the real and imaginary parts of the complex-valued function $f(x) = e^{i\lambda x}$. We obtain (as you should check!)

$$f(M_t) = e^{i\lambda M_t} = f(M_s) + \int_s^t i\lambda e^{i\lambda M_u} dM_u - \frac{1}{2}\int_s^t \lambda^2 e^{i\lambda M_u} du$$

since we are given that $d[M]_u = du$. The integrands are bounded by 1, so that the real and imaginary parts of $\int_s^t i\lambda e^{i\lambda M_u} dM_u$ are square-integrable martingales, and hence $\mathbb{E}\left[\int_s^t i\lambda e^{i\lambda M_u} dM_u | \mathcal{F}_s\right] = 0$. Thus for fixed A in $\mathcal{F}_s$ we multiply the above equation by $1_A e^{-i\lambda M_s}$ and take expectations

$$\int_A e^{i\lambda(M_t - M_s)} dP = P(A) - \frac{1}{2}\lambda^2 \int_s^t \mathbb{E}[e^{i\lambda(M_u - M_s)}] du.$$

This integral equation has the form $g(t) = P(A) - \frac{1}{2}\lambda^2 \int_s^t g(u)du$, with g as the deterministic function $t \to \int_A e^{i\lambda(M_t - M_s)} dP$. Its solution is $g(t) = P(A)e^{-\frac{1}{2}\lambda^2(t-s)}$. This completes the proof, as $A = \Omega$ yields $\mathbb{E}[e^{i\lambda(M_t - M_s)}] = e^{-\frac{1}{2}\lambda^2(t-s)}$, as required.

The alternative proof of the Girsanov Theorem, which we announced in Chapter 7, follows from this result and further use of the Itô formula for martingales. First, we note that, given a probability space $(\Omega, \mathcal{F}, P)$, $[0, T]$ and a probability measure $Q \ll P$ on $(\Omega, \mathcal{F})$, then $M_t = \mathbb{E}\left[\frac{dQ}{dP} | \mathcal{F}_t\right]$ ($t \in [0, T]$) defines a martingale M with the following property:

Lemma 8.20 *The product process* $(X_t M_t)$ *is a P-martingale iff* (X_t) *is a Q-martingale.*

For the proof, simply note that for $s \leq t$ and A in $\mathcal{F}_s$ we have by definition, if XM is a P-martingale

$$\int_A X_t dQ = \int_A X_t M_t dP = \int_A X_s M_s dP = \int_A X_s dQ,$$

so that X is a Q-martingale, while the inner identity shows conversely that XM is a P-martingale.

For our special case of the Girsanov Theorem, we use the martingale $M_t = \exp\left(-cB_t - \frac{1}{2}c^2 t\right)$ to define the measure Q (see Lemma 7.20).

The process M solves the SDE $M_t = 1 - \int_0^t cM_s dB_s$, as is easily verified, and it is strictly positive, so that $Q \sim P$. Being a P-martingale, $\mathbb{E}_P[M_t] = \mathbb{E}_P[M_0] = 1$. So $Q(\Omega) = \mathbb{E}_P[M_T \mathbf{1}] = 1$, and Q is a probability measure.

To show that $X_t = ct + B_t$ defines a Q-BM, we use the Lévy Theorem. Clearly, $X_0 = 0$ and X is path-continuous, so we only need to check that (X_t) and $\left(X_t^2 - t\right)$ are Q-martingales.

The first of these holds iff XM is a P-martingale, by Lemma 8.18. To show that it is, we apply the integration-by-parts formula to XM, noting that $dX_t = dB_t + cdt$, $dM_t = -cM_t dB_t$, so that $d([X, M]_t = -cM_t(dB)_t^2 = -cM_t dt$

$$
\begin{aligned}
d(X_t M_t) &= M_t dX_t + X_t dM_t + d[X, M]_t \\
&= M_t(dB_t + cdt) - cX_t M_t dB_t - cM_t dt \\
&= M_t(1 - cX_t)dB_t.
\end{aligned}
$$

In other words, $X_t M_t = X_0 M_0 + \int_0^t M_s(1 - cX_s)dB_s$ is a stochastic integral, hence a martingale, under P.

For the final condition, apply the Itô formula to $f(x) = x^2$

$$d(X_t)^2 = 2X_t dX_t + d[X]_t = 2X_t dX_t + dt.$$

In other words, $X_t^2 - t = 2\int_0^t X_s dX_s$, and since X is a Q-martingale, so is this stochastic integral. This completes the alternative proof of Lemma 7.20.

The Lévy Theorem shows that BM B plays a central role in the set of continuous $\mathbb{F}$-martingales on the filtered space $(\Omega, \mathcal{F}, \mathbb{F}, P)$ when

$\mathbb{F}$ is the Brownian filtration (i.e. that induced by B). We now outline briefly (following [O]) a key representation theorem for L^2-bounded $\mathbb{F}$-martingales which shows them to be stochastic integrals of B. This, in turn, enables us to fill the remaining gap in our discussion of the Black–Scholes model (Section 8.4) by describing a unique replicating strategy for a European call option in this model. (In fact, the method supplies such strategies for any European derivative security: this confirms that the BS model is *complete*.)

First, we state without proof a lemma which highlights the importance of exponentials already encountered (as the special case where h is constant) in the tilting formula (Section 7.4) and the martingale M used for the Girsanov measure transformation. It shows how Itô integrals of deterministic integrands can create a dense subspace in $L^2(\mathcal{F}_T, P)$.

Lemma 8.21 *The linear span of the set E of random variables*

$$Y_T^h = \exp\left(\int_0^T h(t)dB_t(\omega) - \frac{1}{2}\int_0^T h^2(t)dt\right)$$

for $h \in L^2([0,T], m)$ is dense in $L^2(\mathcal{F}_T, P)$.

For a proof (using analytic extensions and Fourier transforms), see [O].

Theorem 8.22 *(Itô Representation Theorem) If $X \in L^2(\mathcal{F}, P)$, then there is a unique $H \in \mathcal{M}_{[0,T]}^2$ such that*

$$X(\omega) = \mathbb{E}[X] + \int_0^T H_t(\omega)dB_t(\omega) \; a.s.(P).$$

Proof Suppose $Y_T^h(\omega) = \exp\left\{\int_0^T h(s)dB_s(\omega) - \frac{1}{2}\int_0^T h^2(s)ds\right\}$ for some h in $L^2[0,T]$. Set $Y_t^h(\omega) = \exp\left\{\int_0^t h(s)dB_s(\omega) - \frac{1}{2}\int_0^t h^2(s)ds\right\}$ for $t \in [0,T]$ Applying Itô's formula to Y^h we obtain

$$dY_t^h = Y_t^h\left[h(t)dB_t - \frac{1}{2}h^2(t)dt\right] + \frac{1}{2}Y_t^h h^2(t)dt = Y_t^h h(t)dB_t,$$

which means that $Y_t^h = 1 + \int_0^t Y_s^h h(s)dB_s$ for each t in $[0,T]$. This shows that the theorem holds when $X = Y_T^h$ and thus for any Z in the subspace E generated by the Y_T^h. For general X in $L^2(\mathcal{F}, P)$, we use approximation by functions in this dense subspace: suppose $Z_n \to X$ in L^2-norm and $Z_n(\omega) = \mathbb{E}[Z_n] + \int_0^T h_n(s, \omega) dB_s(\omega)$ for some h_n in $\mathcal{M}_{[0,T]}^2$. The Itô isometry yields, for any m, n

$$|||Z_m - Z_n||_2^2 = \mathbb{E}\left[\mathbb{E}[(Z_m - Z_n)] + \left(\int_0^T \{h_m(s) - h_n(s)\} dB \right)^2 \right]$$

$$= \mathbb{E}[(Z_m - Z_n)] + \int_0^T \mathbb{E}[(h_m(s) - h_n(s))^2] ds.$$

The LHS goes to 0 as $m, n \to \infty$, hence we see that $(h_n)_n$ is Cauchy, hence convergent, in $L^2([0, T] \times \Omega)$. The limit H is again in $\mathcal{M}_{[0,T]}^2$, since we have $a.s.(P)$-convergence for a subsequence, hence $H(t, \cdot)$ is $\mathcal{F}_t$-measurable for almost all t, so that H has an adapted version. Using the Itô isometry again, we see that

$$X = \lim_n Z_n = \lim_n \left(\mathbb{E}[Z_n] + \int_0^T h_n(s) dB_s \right)$$

$$= \mathbb{E}[X] + \int_0^T H(s) dB_s,$$

where the limit is taken in $L^2(\mathcal{F}, P)$. Uniqueness follows from yet another application of the Itô isometry.

Corollary 8.23 *If M is an $L^2(\mathbb{F}, P)$-martingale, then there exists a unique H in $\mathcal{M}_{[0,T]}^2$ such that $a.s.(P)$ for all t in $[0, T]$*

$$M_t = \mathbb{E}[M_0) + \int_0^t H_s dB_s.$$

This follows at once from the fact that M is u.i., so that for each $t \leq T$, $M_t = \mathbb{E}[M_T | \mathcal{F}_t]$ and $M_T \in L^2(\mathcal{F}, P)$, while the stochastic integral is a martingale.

Example 8.24 Finally, let us return to the Black–Scholes option pricing model. Under the risk-neutral measure Q, the process B^* with $B_t^* = B_t + \frac{\mu-r}{\sigma} t$ is a BM, $d\tilde{S}_t = \sigma a_u dB_t^*$, so that for a replicating strategy $(a_t, b_t)_{t \in [0,T]}$ the discounted value process $\tilde{V}$ is given by the stochastic integral

$$\tilde{V}_t = V_0 + \int_0^t \sigma a_u \tilde{S}_u dB_u^*.$$

The payoff of a European call with strike K is $f_T = (S_T - K)^+$. The discounted final value $\widetilde{V}_T = a_T \widetilde{S}_T + b_T$ of such a strategy must equal the discounted payoff, i.e. $\widetilde{V}_T = e^{-rT}(S_T - K)^+$. But $\widetilde{V}$ is a u.i. martingale, so that $\widetilde{V}_t = \mathbb{E}_Q[e^{-rT}f_T|\mathcal{F}_t]$. This is an $L^2(\mathcal{F}, Q)$-martingale, so by Corollary 8.22 we can find H in $\mathcal{M}^2_{[0,T]}$ such that for each t, $\widetilde{V}_t = \mathbb{E}[e^{-rT}f_T] + \int_0^t H_u dB_u^*$. Equate the two expressions for $\widetilde{V}_t$ and recall that $V_0 = \mathbb{E}[e^{-rT}f_T]$. We see that $H_t = \sigma a_t e^{-rt}S_t$, as the integrand in the Martingale Representation Theorem is unique (as a element of $\mathcal{M}^2_{[0,T]}$). In other words, the unique stock holding strategy, which replicates the call option, is given by $a_t = \frac{e^{rt}}{\sigma S_t} H_t$. The bond holding is then given by $b_t = \widetilde{V}_t - a_t \widetilde{S}_t = \widetilde{V}_t - \sigma^{-1} H_t$. We have verified the existence of a unique replicating strategy, whose values are determined fully by the integrand H in the martingale representation of the discounted value process. This justifies the assumption we made in deriving the Black–Scholes formula.

References

[A] R.B. Ash, *Real Analysis and Probability*, Academic Press, New York, 1972.

[B-Z] Z. Brzezniak and T. Zastawniak, *Basic Stochastic Processes*, Springer, London, 1998.

[C-K] M. Capinski and E. Kopp, *Measure, Integral and Probability*, 2nd edition, Springer, London, 2004.

[D-M] C. Dellacherie and P.-A. Méyer, *Probabilities and Potential*, North-Holland, Amsterdam, 1975, 1982.

[E-K] R.J. Elliott and P.E. Kopp, *Mathematics of Financial Markets*, 2nd edition, Springer, New York, 2005.

[K-S] I. Karatzas and S.E. Shreve, *Brownian Motion and Stochastic Calculus*, 2nd edition, Springer, New York, 1991.

[K] P.E. Kopp, *Martingales and Stochastic Integrals*, Cambridge University Press, 1984.

[L] M. Loeve, *Probability Theory I, II*, 4th edition, Springer, New York, 1977, 1978.

[N] J. Neveu, *Discrete-Parameter Martingales*, North-Holland, Amsterdam, 1975.

[O] B. Oksendal, *Stochastic Differential Equations*, 6th edition Springer, Heidelberg, 2003.

[P] P. Protter, *Stochastic Integration and Differential Equations*, Springer, New York, 1990.

[R-Y] D. Revuz and M. Yor, *Continuous Martingales and Brownian Motion*, Springer, New York, 1991.

[R-W] L.C.G. Rogers and D. Williams, *Diffusions, Markov Processes, and Martingales*, Cambridge University Press, 2000.

[R] W. Rudin, *Real and Complex Analysis*, McGraw-Hill, New York, 1966.

[S] J.M. Steele, *Stochastic Calculus and Financial Applications*, Springer, New York, 2000.

[W] D. Williams, *Probability with Martingales*, Cambridge University Press, 1991.

Index